STRATEGIC STUDIES INSTITUTE

The Strategic Studies Institute (SSI) is part of the U.S. Army War College and is the strategic-level study agent for issues related to national security and military strategy with emphasis on geostrategic analysis.

The mission of SSI is to use independent analysis to conduct strategic studies that develop policy recommendations on:

- Strategy, planning, and policy for joint and combined employment of military forces;
- Regional strategic appraisals;
- The nature of land warfare;
- Matters affecting the Army's future;
- The concepts, philosophy, and theory of strategy; and,
- Other issues of importance to the leadership of the Army.

Studies produced by civilian and military analysts concern topics having strategic implications for the Army, the Department of Defense, and the larger national security community.

In addition to its studies, SSI publishes special reports on topics of special or immediate interest. These include edited proceedings of conferences and topically oriented roundtables, expanded trip reports, and quick-reaction responses to senior Army leaders.

The Institute provides a valuable analytical capability within the Army to address strategic and other issues in support of Army participation in national security policy formulation.

Strategic Studies Institute
and
U.S. Army War College Press

STAND UP AND FIGHT! THE CREATION OF U.S. SECURITY ORGANIZATIONS, 1942-2005

Ty Seidule
Jacqueline Whitt
Editors

April 2015

*****

Comments pertaining to this report are invited and should be forwarded to: Director, Strategic Studies Institute and U.S. Army War College Press, U.S. Army War College, 47 Ashburn Drive, Carlisle, PA 17013-5010.

*****

All Strategic Studies Institute (SSI) and U.S. Army War College (USAWC) Press publications may be downloaded free of charge from the SSI website. Hard copies of this report may also be obtained free of charge while supplies last by placing an order on the SSI website. SSI publications may be quoted or reprinted in part or in full with permission and appropriate credit given to the U.S. Army Strategic Studies Institute and U.S. Army War College Press, U.S. Army War College, Carlisle, PA. Contact SSI by visiting our website at the following address: *www.StrategicStudiesInstitute.army.mil.*

*****

The Strategic Studies Institute and U.S. Army War College Press publishes a monthly email newsletter to update the national security community on the research of our analysts, recent and forthcoming publications, and upcoming conferences sponsored by the Institute. Each newsletter also provides a strategic commentary by one of our research analysts. If you are interested in receiving this newsletter, please subscribe on the SSI website at *www.StrategicStudiesInstitute.army.mil/newsletter.*

ISBN 1-58487-678-6

CONTENTS

FOREWORD

From the National Security Act of 1947 to the intelligence and security restructuring after September 11, 2001 (9/11), our nation has stood up new security organizations to meet new challenges. On October 10, 2010, my team and I helped give life to a new U.S. security organization when I assumed command of the United States Army Cyber Command. As its first commander, I was more than "present at the creation;" I was responsible for it. In a crowded, yet resource-constrained defense establishment, Army Cyber was responsible for all Army cyber efforts—not only in relation to U.S. Cyber Command, but also the Department of Defense and U.S. Intelligence Community.

A single-page General Order from the Secretary of the Army served as the command's founding document and mission statement. Army Cyber Command was designated the lead for Army missions, actions, and functions related to cyberspace.[1] This included planning, coordinating, and directing the operations and defense of all Army networks, and when directed, conducting full-spectrum cyberspace operations to ensure our own freedom of action in cyberspace and to deny the same to our adversaries. The overall mission was clear enough, but broad and generally in uncharted territory when it came to specifics. Much like the virtual world we live in, this new operational domain of cyberspace was immature and continuously evolving. All aspects required clarity, precision, and focus to ensure unity of effort.

The lack of clarity in the task was not the only challenge—we needed to build the right team to accomplish this new mission. Indeed, people are the centerpiece to all organizations, but human nature is often resistant to change. Thus, transformational change requires effective leadership to excite and motivate the workforce to common purpose. Assembling a team of elite, trusted, competent cyber professionals of character who are excited and committed to the vision and mission is essential in executing the mission daily, while setting the future direction. As is the case with many security organizations, this was top to bottom transformational change.

While transformational change often starts with changes in law, statute, or funding streams, establishing a new operational command focuses on mission, vision, and leadership. The ability to be a part of transformational change, whether it is legislated or operational, is rare and unique. Assuming command of Army Cyber presented a once-in-a-lifetime opportunity to help make a difference in an area critical to national security: Cyber threats were growing rapidly, and attempts to penetrate our networks were increasing in frequency and complexity.

While we knew the threats were evolving, in the beginning, there were more unknowns than knowns. In fact, there was little agreement or understanding of terms and definitions. No doctrine or policy for cyberspace existed, and the existing culture took our freedom to operate in cyberspace for granted. Establishing cyberspace as an operational domain was a necessary change in order to meet the security challenges our Nation faces in cyberspace.

In standing up a security organization, leaders must consider historical lessons early on—before a crisis point occurs. As it had been a long time since a new Army-level command was established, I asked the Department of History at my alma mater, The United States Military Academy at West Point, to provide some lessons learned about standing up a new security organization. Unsurprisingly, Colonel Ty Seidule was already considering these issues. He and his team provided a historically grounded appreciation of what it means to stand up a new security organization. These lessons became the basis of this fine manuscript and were invaluable to me as we grew and matured the command. All were relevant during Army Cyber Command's first three years and several affected our daily thought and action. Early discussions about these security lessons learned allowed me to consider the potential friction of organizational rivalry, the need to change a culture, the caution to avoid bad analogies, the requirement to develop simulations, and the importance of allies.

Arguably, an appreciation of historical perspective and context is even more important in blossoming agencies, when the contours of the agency are coalescing around a mold tempered by many—often-competing—inputs. Early decisions are critical, since they often map and shape the command as it develops. Indeed, these early decisions (or what might be called in the cyber field, "founding frequencies") often matter more than strategic decisions in well-established agencies.

Security organizations stand up in response to external, often traumatic, events and pressures, cultural or geopolitical trends, the search for fiscal efficiencies, or changes in technology. These new missions demand that new organizations deal with a significantly different landscape than what has come before. The speed at which cyber threats were growing required a sense of urgency to increase our capacity and capability to conduct cyber operations. The margin for error was shrinking, and we needed to get it close the first time, while remaining agile and adaptive. People often think the Army is a big, hidebound bureaucracy, but actually the Army is ever evolving and changing.

For much of the Cold War, new security organizations were often established in support of new mission sets, new potential enemies, and they frequently involved breaking organizations into their constituent parts. However, after 9/11, new agencies could emerge by consolidating several agencies, for example, creating The Department of Homeland Security or merging U.S. Space Command with U.S. Strategic Command, not just by hiving off mission sets, such as the separation of the U.S. Air Force from the Army Air Corps. In standing up new security organizations, the mission needs to be relevant and add value to national security. If the American people do not see its necessity, the organization cannot endure.

In this groundbreaking edited volume, West Point historian Colonel Seidule and Air War College strategist Dr. Jacqueline E. Whitt provide a much needed framework for conceptualizing the development and take off of new security organizations. They—and the chapter authors—illustrate the relevance of historical lessons for current strategy. Their approach encourages us to learn from the past, consider proper historical parallels, and identify previous challenges and solutions.

This volume should be required reading for policymakers, military officers, and students of American and international history, but anyone charged with transformational change will find applicable historical lessons in this volume. The history of establishing new security organizations is not for historians only, but also for practitioners and leaders who find themselves in an environment of continuous change and transformation, and who are charged to Stand Up and Fight!

Rhett A. Hernandez
Lieutenant General, U.S. Army, Retired
West Point Distinguished Cyber Chair,
Army Cyber Institute

ENDNOTE

1. Department of the Army General Order 2010-26, "Establishment of the U.S. Army Cyber Command," U.S. Army Cyber Command Historical Files, October 1, 2010.

PREFACE

Starting a new security organization is a difficult business. Hundreds of questions come in staccato bursts; each requires a decision and each decision takes on the permanence of tradition. Tradition becomes culture and a new organization becomes the sum of those early decisions. In this collection of essays, historians, most of them military officers, try to grapple with the challenges of creating new security organizations. Our aim is to help those few men and women who start new governmental bodies charged with protecting the American people to make sound and historically informed decisions by highlighting several common themes for consideration.

We began this project in early-2011 when Colonel Greg Conti of West Point's Information Technology Operations Center asked us to assist the fledgling Cyber Command (USCYBERCOM) as it sprang into existence under the command of General Keith Alexander. A series of cyber attacks around the world had galvanized Congress and the Department of Defense to act. Colonel Conti, working for General Alexander, wondered if historians might offer useful perspectives as USCYBERCOM made quick decisions about its future.

In response to Colonel Conti's request, 14 historians at West Point, NY, volunteered to look at a variety of security organizations created since World War II. We looked at functional commands and regional commands, commands that thrived and commands that died, commands created in response to technological changes and those created during bursts of reform. We chose a wide variety of different commands to allow for the broadest view possible.

Our analysis produced no easy answers on how to create an effective, efficient, stable and politically viable security organization. There proved to be no readymade, "shake and bake" formula. Smart people made the best decisions they could with limited information and time. Often, those decisions helped create great organizations. Of course, some decisions led to failure or even catastrophe.

If we can provide no pat answers, why even try? This question is a fundamental concern for historians who are interested in speaking to contemporary policy and policymakers. Most of us involved with the project called West Point's Department of History home as the project was conceived and as chapters were written and edited. There, in the bowels of Thayer Hall, surrounded by the inescapable reminders of the institution's past, as we taught young cadets to think and write critically about the past, we strove also to be loyal to our motto, *Sapientia per Historiam* (Wisdom through History), as scholars. We firmly believe history can help define the right questions to ask and provide context about how sister organizations function. History, then, can broaden one's knowledge. An old aphorism is that the thing most like war is other war. Similarly, the best way to understand one's own organization may be to study another. One of our goals here is to place USCYBERCOM in a larger defense perspective—to understand how agencies and organizations, especially in their formative years, related to other entities—in order to give perspective to the wide range of possibilities and outcomes. Ultimately, this diverse and broadened perspective should alleviate concerns that there is one "right" way, even as it suggests there are perhaps some best practices to follow and pitfalls to avoid.

By looking at the creation of 13 different security organizations over 70 years, we have found issues that warrant attention from every new outfit. If history provides no clear answers, perhaps it can show harried staff officers how others have grappled with similar problems. Perhaps one of these essays will spark a question in the mind of a policymaker, a senior leader, or an "iron major," and lead the organization to see its current problems in a different light. In short, maybe, just maybe, our essays will provide a little wisdom.

CHAPTER 1

INTRODUCTION: A HISTORICAL OVERVIEW OF AMERICAN SECURITY ORGANIZATIONS

Ty Seidule

By 2014, the United States entered the second decade of the "Global War on Terror" —also called "The Long War," or, as Army Chief of Staff from 2007-11 General George Casey named it, "the era of persistent conflict." Facing significant challenges to U.S. security posed by violent extremist organizations, rapidly-changing technologies, and a complex strategic environment, Congress routinely has passed legislation creating or reforming security organizations, assuming, apparently, that better systems and organizations will provide one key to victory. Since World War II, the United States has seen a huge and steady expansion of defense and security organizations tasked to protect the American people from "all enemies foreign and domestic," but this is almost entirely a 20th-century phenomenon. A historical perspective reminds us that the United States went for more than a century without a major reform to the nation's organizational structure for security.

Another major theme from the study of American security policy is the changing meaning of the word "security" itself. As we will see, the first organization created was the Department of War, a cabinet-level department, that handled all military affairs. In 1798, Congress split out control of naval affairs. Thereafter, the Secretary of War controlled the Army. As the name implied, the secretary's job focused on war, and despite the myriad tasks assigned to the Army, from

Indian removal to engineering projects to strike-breaking, security was basically synonymous with war. The Navy, on the other hand, protected commerce. After World War II, with the passage of the National Security Act of 1947, these organizations were unified under the Department of Defense, and its mission and funding increased dramatically.

Since then, the proliferation of organizations created to defend the United States expanded far beyond the confines of distinct Army and Navy missions. However, the national concept of security still focused on external threats. After the September 11, 2001 (9/11) attacks, a new agency was formed, the Department of Homeland Security (DHS). With that change to the national security apparatus, the term "security" became broad enough to cover domestic threats such as homegrown terrorism and natural disasters. As the definition of security has grown to include almost any risk, so too have the resources allocated to it. In the future, if the definition of security continues to grow and if Americans are willing to accept less and less risk, perhaps food, air, and water will also come under control of security organizations.

SECURITY POLICY: 1789-1900

When the states approved the Constitution in 1789, Congress passed legislation to create a Department of War with responsibility for a small land force. When the need arose for a Navy, the Secretary of War took responsibility. In 1798, the demands of running both forces became too great, and Congress created a Department of the Navy on equal footing with the Department of War. The Army and the Navy answered directly to the President. Only in 1947, after 150 years and two world wars, did Congress place the armed services—now including the Air Force—under one cabinet-level secretary. Ultimately, many of the organizations created in the last 60 years stem from the incoherent structures bequeathed by the Founding Fathers.

Congress passed other pieces of security legislation in the 1790s that remained extant for more than a century. The Uniformed Militia Act passed in 1792 set military policy until 1902. It established the dual military tradition of a small Regular Army augmented by militia from the states. Unfortunately, it provided only recommendations to the states with no funding or accountability.[1] When the country went to war in 1812, 1848, 1861, and 1898, the President called for volunteers to bolster the exiguous regulars.[2] Yet, none of these wars led Congress to major security reorganization: the American people and their politicians feared a strong military and kept the Army and the Navy weak during peacetime.

When not at war, the small peacetime security resources went toward continental expansion from the Atlantic to the Pacific Oceans. Those two great oceans protected by the British navy provided security from major European powers as the nation grew westward. Yet despite avoiding major war with Britain, France, and Russia, the American military remained busy in the 19th century, providing security on the frontier and protecting commerce. The Army had the dangerous and unenviable task of evicting Indian tribes from their land and forcing them onto desolate reservations. In some ways, these uses foreshadowed the domestic-security responsibilities of the post-9/11 security apparatus, while still being thought of as a defensive, outward-oriented force.

The services suffered from lack of funding, yet American politicians and military leaders did a better than adequate job of weighing threats and creating a security system that met the needs of the nation. Few of those needs, many politicians felt, required wholesale reform of security laws. The only exception to the paucity of legislation in the 19th century was the creation of intelligence organizations for the Army and the Navy in the 1880s, hardly a precedent for future action.[3]

SECURITY POLICY: 1900-1941

> This is an era of organization.
>
> Theodore Roosevelt[4]

The Spanish-American War, however, changed the security needs of America forever. In the peace treaty that followed that war, the United States gained a fledgling empire with the acquisition of the Philippines, Cuba, Guam, and Puerto Rico. Pacifying and administering these far-flung territories required not only soldiers and sailors but also planners, diplomats, agriculture experts, doctors, and engineers, in short a security system far more sophisticated than that created in the 1790s.

To create the new system, President William McKinley wisely chose Elihu Root, the Secretary of War from 1899 to 1904. A corporate lawyer with no military experience, Root was an astute politician. Through shrewd maneuvering aided by like-minded political and military leaders, Root crafted a series of laws that established the beginnings of an overarching American security establishment. It set the tone for all subsequent laws aimed at creating a more effective defense system. True to the goals of contemporary Progressivism, a movement that aimed not only to organize and professionalize America, but also improve its moral fiber, Root tackled the problem of rendering the peacetime army capable of fighting a war through the notion of reform.[5] By 1903, among other things, Root had established a General Staff, reduced the influence of bureau chiefs, reorganized the Militia into the National Guard and Reserve Militia, reorganized and expanded the Medical Department, created the Quartermaster Corps, and established the U.S. Army War College. Following these so-called "Root Reforms," each future attempt to create a better, more effective, and more efficient organization would be billed as reform. Perhaps ironically, the notion of reform, promoted by Progressives as a means of overcoming the natural selfishness of man to create better relations between all humans, had been harnessed to the end of organizing more efficiently for war.[6]

Seen in this way, each security reorganization since the early part of the last century was an attempt to create not merely effectiveness or efficiency, but harmony. No wonder the expectations of new organizations are so high, and why they seldom reach their stated goals. When they fail to achieve the unachievable, Americans call for more reform, creating a cycle that continues to this day.

Further reform legislation for both the Army and the Navy followed the Root Reforms. World War I led Congress to pass National Defense Acts in 1916 and 1920. The word "defense" here was crucial. For the first time, military policy in peacetime included more than the Army and the Navy. The United States needed all the elements of national power to fight a world war. The legislation in 1920, in particular, addressed the need for industry to work with the military services. The first industrial mobilization plan came from the fertile mind of Dwight D. Eisenhower, a young Army staff officer.[7] The beginnings of interagency organizations came in the plans drawn up to harness the American economy to fight.

World War I also highlighted the importance of technology in combat, particularly manned flight. Arguments over who would control the air eventually spilt the Army into two services and involved a fight with the Navy as well. Before 1941, the Army-Navy joint board met regularly to adjudicate disputes; however, those disputes were rarely resolved, unless the issue reached the President or resulted in legislation.

SECURITY POLICY DURING WORLD WAR II

> It is a war of smokestacks as well as of men.
>
> George C. Marshall[8]

During World War II, it is only a slight exaggeration to say that the United States, at least initially, found combined operations between countries easier to execute than joint operations between services.[9] In discussing strategy with their allies, the Americans were at a disadvantage because they had no organization comparable to the British Joint Chiefs and tended to argue among themselves at strategic conferences instead of presenting a unified front to the British. Eventually, in 1942, the Americans created a joint staff and became more adept at presenting strategic alternatives to the British.

Early in the war, battlefield failure, competition for resources, and problematic command structures led to intense rivalry among the services. Problems in the Pacific Theater were so acute that the Joint Chiefs in Washington had to make all major decisions, yet by 1944, in all theaters of the war, the services had worked through their problems and were acting mostly in concert. By D-Day, General Dwight D. Eisenhower had finally convinced the Army Air Corps to conduct air interdiction missions, and by the summer of 1944 even bombers were used in a close air support role (albeit with sometimes regrettable results).[10] Likewise in the Pacific, the initial rivalry between the Army and the Navy was resolved as both services supported the dual thrusts of Admiral Chester Nimitz in the Central Pacific and General Douglas MacArthur in the Southwest Pacific. Of course, unlimited funds and extraordinary resources pumped out of the American economy and into all services in 1944 and 1945 lubricated the joint engine of war.

World War II showed clearly that organization mattered. After the debacle of the Pearl Harbor, HI, raid, intelligence agencies broke military and diplomatic codes and then shared them with senior military leaders in a timely fashion. The military also reorganized at all levels and created processes to take battlefield lessons learned after early debacles against the Japanese in Buna, New Guinea, and against the Germans at the Kasserine Pass, Tunisia, and incorporate them into strategy, operations, and tactics. The final amphibious operations in Okinawa, Japan, and Normandy, France, were among the best organized campaigns in the history of warfare. The creation of joint and combined staffs helped articulate goals and set priorities. Ultimately, the Allies won the global war.

One of the many reasons they ultimately triumphed was that the American government, the military, and industry proved themselves to be far better planners and organizers than their German and Japanese counterparts.[11]

Victory came, in part, because the United States and its allies out-organized and out-produced the Axis powers. Americans built tens of thousands of *Liberty* ships, B-29 bombers, Sherman tanks, jeeps, and two atom bombs through hard work, ingenuity, and, yes, organization. Fighting a world war on multiple fronts against several different enemies required more than just military services. America integrated military and industry to a degree never seen before. The only way to marry two forces with such different ethos and aims was through superior organization. Assigning priority for thousands of commodities like nylon, cotton, steel, leather, and rubber required centralized planning. Which were more important, airplanes, tanks, or carriers? Those disputes need adjudication at the national level and then each service had to prioritize. Should fighters or bombers receive priority? Tanks or half-tracks? At every level, war required deep institutional knowledge and organization.

Inventing, then producing and then implementing technological innovations also required organization. The best example was the Manhattan Project that created the atom bomb. Technology required a symbiotic relationship between industry and government. In World War II, the government led the technological charge. Industry alone had neither the need nor the ability to harness the resources to create an atomic bomb in such a radically short time. Today, that trend is often reversed and industry creates and shapes technology forcing the military to react.

SECURITY POLICY SINCE 1945

> The military profession organizes men so as to overcome their inherent fears and failings.
>
> Samuel Huntington[12]

In 1947, Congress passed the most important security legislation since the Root Reforms, the National Security Act, setting up the basic contours of all subsequent security legislation.[13] Its goal was to solve two major problems. First, with the creation of the Central Intelligence Agency, the Act meant specifically to prevent another Pearl Harbor. The solution to the second problem, interservice bickering, was the creation of a Secretary of Defense in the chain of command over the services. When the initial law provided a secretary too feeble to exercise real power, Congress strengthened the position in a series of amendments in 1949 and 1950. The trend throughout the post-war era saw ever greater involvement of civilians in military policy and in areas once thought of as strictly concerning the profession of arms. One example was the Uniform Code of Military Justice passed by Congress in 1950, that took power away from military commanders and instituted a civilian "Court of Military Appeals" over the courts martial system. The services had only themselves to blame. The Army, Navy and Air Force could not agree on roles and missions or allocation of the defense budget. The interservice fights became more public, more pronounced, and more vicious as each branch used members of Congress and the press to help fight their turf battles. The early post-war period set the tone for strong civilian leadership and tough service fights.[14]

After 1947, the next major restructuring came in 1986 with the Goldwater-Nichols Department of Defense Restructuring Act that addressed sacred service interests—personnel and procurement.[15] Services that had worked well together in fighting a world war with relatively unlimited resources proved to be poor teammates in a resource-constrained environment. The 1986 Act aimed to fix the interservice tensions stemming from defeat in the Vietnam War, the failed Iranian hostage rescue of 1980, and the successful, but awkwardly executed, invasion of Grenada in 1983. The Goldwater-Nichols Act took power away from the services and gave it to regional commanders who no longer had to plead their case to the Army, Navy, and Air Force chiefs. It also forced the services to start a process of shared procurement, especially for cutting edge technology. Additionally, Congress created an all-services staff composed of officers destined one day for flag rank. Then Congress forced the services to create a new education system to prepare officers to work together as a "joint staff." Finally, Congress gave the Chairman explicit remit to recommend changes in roles and mission for the services, particularly in regards to new technology.

The third and final major security legislation in the post-war era occurred in 2002 with the Homeland Security Act. In response to the al-Qaeda terrorist attacks in 2001, the President and Congress created the DHS. Congress clearly articulated its broad mission, "Prevent terrorist attacks within the United States."[16] The same reforming drive that created the 1947 National Defense Act to prevent another Pearl Harbor animated the political process after the 9/11 attacks on the World Trade Centers in New York and the Pentagon in Washington, DC. Government, and lots of it, was the best way to secure the nation from future terrorist attacks. DHS had its start in an era of crisis that prevented rational thought about the best way to accomplish its assignment.

The enormous DHS mission expanded after Hurricane Katrina struck the U.S. Gulf Coast in 2005. In addition to preventing terrorist attacks, it would also lead the nation's response to natural disasters. Yet DHS has no military or paramilitary arm capable of accomplishing the missions it was assigned. Despite several reorganizations, the initial problems of the organization have gone unresolved. Although a thorough review of the first 10 years might provide some valuable insights, DHS is the only agency in this study that has no official historian assigned to it. Nonetheless, because of its gargantuan size and scope, all other agencies will have to find a way to deal with it over the next decade.

Organizations and resources dedicated to security have shown steady growth, particularly after 1945. The three biggest changes since 1945 (the 1947 National Security Act, 1986 Goldwater-Nichols Act, and the 2002 Homeland Security Act) have come as a result of attack or defeat, and each act expanded the concept of security. The word "security" has proved remarkably elastic, encompassing threats that a previous generation would have seen as problems outside the remit of security organizations. The security industry shows no sign of slowing.

All professionals look to expand their organizations, and the security specialists have been remarkably successful since 2001. The professionalization of security is in full swing. Budding security specialists can now receive a master's and doctorate degrees in homeland security in the same way that the post-

1945 generation developed security studies programs to counter the Soviet nuclear threat. San Diego State University (San Diego, CA) and the University of Colorado-Colorado Springs (Colorado Springs, CO), among others, offer doctorate degrees in homeland security along with a host of online schools. Should the security studies programs at prestigious universities like Stanford (Stanford, CA) and Johns Hopkins (Baltimore, MD) think of these programs dismissively, they should remember that established fields like history and political science initially looked at their programs as unworthy disciplines in the 1940s and 1950s.

ENDNOTES - CHAPTER 1

1. Allan R. Millett and Peter Maslowski, *For the Common Defense: A Military History of the United States of America*, Rev. and Expanded Ed., New York: The New Press, 1994, pp. 92-94.

2. Only during the Civil War did America resort to a draft in the 19th century. In that case, the draft often served as more of an inducement to enlist than an actual form of manpower mobilization.

3. Millett and Maslowski, p. 102.

4. Theodore Roosevelt, "It Takes More than that to Kill a Bull Moose," speech given in Milwaukee, WI, October 14, 1912. Before delivering his remarks, Roosevelt was shot by John Shrank. Instead of going immediately to the hospital, Roosevelt delivered the speech with the bullet still in his body. See Speeches, Theodore Roosevelt Association, available from *www.theodoreroosevelt.org/research/speech%20kill%20moose.htm*.

5. Millett and Maslowski, pp. 326-332.

6. See Samuel Haber, *Efficiency and Uplift: Scientific Management in the Progressive Era 1890-1920*, Chicago, IL: The University of Chicago Press, 1964.

7. See Ty Seidule, "Eisenhower in the Interwar Years," Francois Cochet, ed., *De Gaulle et les "Jeunes Turcs" dans les armées occidentals (1930-1945): Une génération de la réflexion à l'action* (*DeGaulle and the "Young Turks" in the Western Armies (1930-1945): A Generation of Reflection to Action*), Paris, France: Actes academiques, 2008.

8. George C. Marshall, "Address to the Chamber of Commerce of the United States," April 29, 1941.

9. Combined operations are between the services of two or more nations. Joint operations involve the services (Army, Navy, and Air Force) of one nation. Combined arms involve branches from one service such as infantry, armor, and artillery.

10. Operation COBRA successfully used bombers to break out from the Cotentin Peninsula during the Normandy campaign. However, the first bombings struck Allied soldiers and caused thousands of casualties, including Lieutenant General Leslie McNair, the highest ranking American officer killed during World War II.

11. Richard Overy, *Why the Allies Won*, New York: Norton, 1997.

12. Samuel Huntington, *The Soldier and the State*, Cambridge, MA: Belknap, 1981, p. iii.

13. Douglas T. Stewart, *Creating the National Security State: A History of the Law That Transformed America*, Princeton, NJ: Princeton University Press, 2011.

14. Millett and Maslowski, pp. 504-506.

15. See *www.au.af.mil/au/awc/awcgate/congress/title_10.htm*.

16. See *www.gpo.gov/fdsys/pkg/CRPT-107hrpt609/pdf/CRPT-107hrpt609-pt1.pdf*.

PART I:

UNITED STATES UNIFIED COMBATANT COMMANDS

CHAPTER 2

LEGISLATING CHANGE: THE FORMATION OF U.S. SPECIAL OPERATIONS COMMAND

Brian P. Dunn

INTRODUCTION

Covered in great detail by Susan Marquis and William Boykin in respective works and discussed later in this chapter, the formation of the U.S. Special Operations Command (USSOCOM) and the road to achieving and sustaining a joint special operations forces (SOF) capability was tortuous. Ironically, the very distaste of the Department of Defense (DoD) for SOF in the post-Vietnam era and the intransigence of the Pentagon, Washington, DC, to reform resulted in the bureaucratically radical formation of USSOCOM. USSOCOM's structure and orientation were, in fact, not driven by the unmistakable military necessity of low-intensity conflict, but by the very resistance of the DoD to both acknowledge and address this necessity. Reformers from outside of the organization, recognizing both the Pentagon's uncorrected focus on the conventional possibilities of the Cold War and its dogmatic insistence on conventional strategy and procuring, at any cost, the weapons programs needed to win a conventional fight, rallied to the side of SOF employment. Offering SOF the effective advocacy previously denied them by the DoD, the fate of an entire unified combatant command would ultimately rest largely with several legislators.

Commanding over 80,000 active-duty service members and managing a budget exceeding $6 billion for the fiscal year beginning in 2010, USSOCOM is not only a key tool in the nation's national security apparatus, but is also an entrenched institution within an admittedly bureaucratic DoD.[1] Impressively, USSOCOM is also a relatively new institution whose current structure, clout, and cache offer little hint of the painful struggle behind its inception. Far from being born primarily from the organizational necessities of warfare, the story of USSOCOM is, instead, one of intense bureaucratic infighting, charged disagreements about the nature of warfare in the late-Cold War, and challenges to the constitutional fundamentals of civilian control over the military.

Although officially operational in April 1987, the story of USSOCOM is rooted firmly in Southeast Asia, where a wide variety of special operations were employed to support the American mission in Vietnam, Cambodia, and Laos.[2] Indeed, U.S. Army Special Forces were themselves a product of President John F. Kennedy's attempt to "stop the North Vietnamese aggression without getting American combat troops heavily involved in the war."[3] Obviously unsuccessful in achieving this lofty goal, special forces continued to operate in parallel or conjunction with conventional U.S. forces as part of the mission in Vietnam. As U.S. military involvement in the conflict wound down and U.S. interest in protracted nonconventional conflicts understandably decreased, SOF were one of the first bill payers for the post-Vietnam army. Indeed, while the DoD budget sustained relatively modest decreases as a portion of total domestic discretionary spending (slipping from 59.1 percent in 1973 to 48.7 percent in 1979) and sustained itself year-to-year in terms of real dollars,[4] SOF funding plunged by roughly 95 percent. This translated into a 66 percent reduction in the size of the Army's Special Forces groups, a massive reduction in SOF-capable aircraft, and the decommissioning of the Navy's sole SOF-capable submarine.[5] Viewed largely by the DoD as an early and somewhat misguided experiment in handling the conflict in Vietnam, the end of American involvement in Southeast Asia rendered unnecessary the wide-scale retention of special operations forces.

Other conventional attitudes and apprehensions arrayed against the sustainment of designated SOF units in the post-Vietnam era, with many of these perceptions hindering the eventual formation of USSOCOM well over a decade later. Even during the height of the conflict in Vietnam, there existed significant "mistrust, suspicion, and lack of understanding" between conventional and unconventional forces, where SOF personnel and units were deemed, with some degree of accuracy, to be "secretive, elitist," and "clannish."[6] Senior uniformed leaders, whose reaction to the inception of the Army's "Green Beret" program in the early-1960s had been "overwhelmingly negative," found little to like about the unconventional forces as the war appeared increasingly unwinnable.[7] With the eventual withdrawal of U.S. forces, this lasting distaste for the unconventional war in Vietnam, coupled with the existing distrust of unconventional forces, left many disinclined to champion the retention of SOF in the post-war period. This particular variant of the Vietnam Syndrome proved persistent enough to complicate the arguments surrounding the institutionalization of USSOCOM years later.

Eager to jettison the albatross of Vietnam, the DoD, as an institution, reoriented itself away from SOF-oriented missions even before direct military involvement in the conflict in Vietnam had ceased. Indeed, in his 1970 report to Congress entitled "A Strategy for Peace: A National Security Strategy of Realistic Deterrence," President Richard M. Nixon's National Security Council laid out a one-and-a-half war concept designed to fight a major conventional war in Europe while also fighting in a "sub-theater" conflict (with the Middle East the likely locale). Supporting this conventionally engineered initiative, National Security Decision Memoranda 95 deemed it "vital that NATO [the North Atlantic Treaty Organization] have a credible conventional defense posture to deter and, if necessary, defend against conventional attack by Warsaw Pact forces."[8] The services, with the Army in the lead, were "quick to embrace this return to a more conducive and comfortable strategic environment."[9] Concurrent to this conventional reorientation, Noel C. Koch, a deputy Secretary of Defense, concluded there was a "total absence of defense policy on SOF during the 1970s."[10] To the extreme detriment of SOF, this orientation towards conventional warfare in Europe became increasingly ingrained in the institutional thinking of DoD planners and, more importantly, the Service Chiefs.

This reorientation towards preparing for a conventional showdown with the Soviet Union drove major (and expensive) platform initiatives for each of the services. Taking over as the Army Chief of Staff from his previous position as the head of Military Assistance Command in Vietnam, General Creighton Abrams lost no time in promoting his "Big Five," which would eventually produce some of the mainstays of the Army's arsenal: a main battle tank (eventually the M1 Abrams), infantry fighting vehicle (the M2/3 Bradley), attack helicopter (the AH-64 Apache), utility helicopter (the UH-60 Blackhawk), and a surface-to-air missile (the Patriot system).[11] The Navy likewise took cues from the reorientation towards conventional conflict, initiating programs for guided missile cruisers, *Nimitz*-class aircraft carriers, *Ohio*-class ballistic missile submarines, *Los Angeles*-class submarines, F-14 Tomcats, and F/A-18 Hornets.[12] The Air Force's initiation of programs promoting the C5 Galaxy transport, F16 Fighting Falcon, and F15 Eagle demonstrated a similar interest in matching strategic policy with the hardware to fight and win conventional wars.

These major program initiatives, especially when born in a period of constrained budgeting, gave rise to significant interservice competition as proponents struggled over increasingly limited funding. Indeed, Henry Kissinger predicted as much in 1968 when he noted that "given the likelihood of continuing limits on defense spending . . . there will be intense competition among the Military services for the limited resources," that "could lead to a return of the interservice battles of the 1950s and overwhelm any rational defense planning."[13] Spotlighting such a potential shift towards "irrational" warfighting by the early-1980s was both the feudal nature of interservice interaction and the atrophied state of the U.S. special operations arsenal and capabilities. In addition to a spate of terrorist attacks and hijackings in the late-1970s, Operation EAGLE CLAW, the failed hostage rescue attempt in Iran, and Operation URGENT FURY, the invasion of Grenada, triggered a substantial reevaluation of how special operations forces from the various services would fit into the conventional DoD. Incredibly, the valuable lessons garnered from both events were insufficient to spur internal change in a DoD increasingly addled by the interservice fighting Kissinger had predicted. Instead, the inadequate response by the DoD and its services eventually generated the extraordinary external pressure capable of addressing fundamental problems regarding the organization and application of SOF in the DoD.

No single event was more critical to the eventual revitalization and reorganization of special operations forces than the failed attempt to rescue the American hostages in Iran. As a tragic capstone to the prolonged captivity of American hostages, the aborted mission was a significant blow for not only the special operations communities in the participating services, but for national morale. In the wake of the joint mission, a high-profile, publicized report by Admiral James Holloway observed that, given the complexity of the mission, there was insufficient interoperability between the SOF components from the four donor services. According to Holloway, this deficiency was driven by the very nature of the divided missions and fiefdoms within the DoD. Rather than a standing organization that remained perpetually capable of accomplishing such a challenging mission, the report stated:

> [t]he Joint Chiefs of Staff had to start, literally, from the beginning to establish a JTF [Joint Task Force], create an organization, provide a staff, develop a plan, select the units, and train the force before the first mission capability could be attained,[14]

meaning that the operation was, by the very structure of DoD, "compartmentalized and reliant on ad hoc arrangements."[15] Coupled with a recommendation to form a permanent "Special Operations Advisory Panel" to better inform planners and policymakers of the capabilities and limitations of unconventional forces, the report also suggested that "an existing JTF organization . . . would have provided an organization framework of professional expertise around which a larger tailored force organization could quickly coalesce." [16] Together, these modest recommendations formed the basis for a complete reorganization of special operations within the DoD.

Although the Pentagon was quick to adopt the Holloway Report's recommendations and form the Counter-Terrorism Joint Task Force and Special Operations Policy Advisory Panel, neither organization contributed significantly towards fully addressing the interoperability of joint SOF.[17] More encouragingly for the long-term benefit of the Special Operations community, the massive failure of the special operations mission in Tehran, Iran, did garner attention both in the halls of Congress and in the Oval Office. Coupled with the spate of international hijackings in the late-1970s and early-1980s and an increased U.S. military presence in Latin America, the failure in Tehran underlined the continued need for special operations. In the age of potential nuclear conflict (according to Senator Sam Nunn, "the least likely . . . and yet admittedly, most awesome of threats")[18] and an unlikely large-scale conventional warfare in Europe, low intensity conflicts (LIC) appeared to many defense experts as "the sort of conflicts that our adversaries [would] resort to in an age of nuclear deterrence." In addition to the small proxy wars taking place across three continents, a renewed Soviet emphasis on *spetsnaz* forces made a U.S. countermove seem self-evident. Secretary of Defense Caspar Weinberger likewise noted that America was "well prepared for the least likely of conflicts and poorly prepared for the most likely," ominously noting the consequences of "failure to deter conflict at the lowest level."[19]

Doctrinally repositioned in the spectrum of U.S. operations, SOF gradually obtained limited obligated funding that modestly bolstered its capabilities. While not addressing the pressing organizational issues facing SOF, congressional funding for special operations increased from .1 percent of the total Defense budget to .3 percent, which, although still meager, represented a full 200 percent jump between 1981 and 1985.[20] Yet while Congress and the administration of Ronald Reagan may have embraced a renewed role for SOF, the DoD did not. Rather than embrace the new exigencies facing the United States, the Service Chiefs retrenched into their four fiefs, reallocated Congressional funding to conventional projects, and spurned congressional oversight.

Indeed, the various services became increasingly territorial about resources, especially as new and increasingly costly programs amplified competition for limited resources. This often put the services at odds with congressional intent, a prime example being the perennial fight between the Air Force and Congress over the purchase of new SOF-capable MC-130 Talons. With a receptive Congress allocating funds each year for the purchase, the Air Force simply ignored legislation in favor of reallocating the resources to big-budget programs driven by the requisites of conventional combat. The Army engaged Congress and the budget process similarly, often under the nose of a disapproving Service chief. After General Edward Meyer publically agreed to partially fund a communications program to assist Army Special Forces, the funding was promptly reprogrammed by his own staff to support a more mainstream initiative.[21] In other circumstances, funds and even physical "equipment tended to get siphoned off" before reaching the SOF user, leading to one instance in which a SOF unit received equipment a full 17 years after it was requested.[22] Not surprisingly, the respective services' behavior often placed them at odds with the hand that fed them. Increasingly frustrated over what representatives felt was a "contravention of law," Congress still found itself largely impotent to control individual SOF program spending within the respective services.[23]

Given that Nixon's one-and-a-half war doctrine was still nominally governing military planning, the program and funding decisions made by the services largely reflected their expectation of a major conventional conflict in Europe with a smaller conventional conflict in another locale. As noted by a Senate Armed Services Committee staff study from the early-1980s, however, this also resulted in a cyclical system where, "the Services have a tendency in force planning to focus on high-intensity conflicts upon which resource programs are justified."[24] The major equipment programs ushered in during the post-Vietnam era were considered essential in the achievement of the services' core functions. Special operations, deemed by most in the DoD to have "never been a core program," largely subsisted on leftovers from the budgeting process.[25]

SOF jockeying for position and money was complicated by the bevy of new defense programs competing for dollars, most notably the imminent and expensive Strategic Defense Initiative, which raised the hackles of Service proponents during the internecine Defense Resources Board. According to one DoD official, the "perennial DoD problem" consisted "of stuffing [10] pounds of program into a five-pound bag." SOF, on the periphery since the end of Vietnam, could expect little in the process, and was further hobbled by conventional wisdom that equated "small with unimportant." While $1 million seemed a drop in the bucket to the massive Defense budgets of the 1980s, such funding was proportionally significant for the funding-starved SOF elements of each service.[26] Representative Dan Daniel, Chairman of the House Armed Services Committee, noted the ultimate success and failure of the budgeting process in that by concentrating on "deterring nuclear conflict and the 'big' war on the plains of Europe . . . we have avoided both. But what we have failed to deter is low-intensity conflict—the peculiar province of SOF."[27] Ironically, the repeated denial of funding to SOF across the services eventually fed the political will to form USSOCOM.

Also critical to the formation of USSOCOM were the lessons learned from the American invasion of Grenada in 1983. The operation highlighted both inadequacies in the employment of SOF and a larger inability for the services to work effectively in a joint environment. Outwardly touted as a U.S. victory, stories of inadequate intelligence, avoidable friendly fire incidents, and a complete breakdown in inter- and even intraservice communication raised serious questions about the U.S. ability to conduct even narrow military missions. The apparent mishandling of SOF during the invasion was also criticized. In testimony before the Senate Armed Services Committee, retired Major General Richard Scholtes, commander of SOF elements for the operation, described his forces as "misused" by conventional commanders because of "numerous fundamental misunderstandings of their tactics and capabilities," that ultimately resulted in the deaths of U.S. servicemen. Ranking Senators William Cohen of Maine and Sam Nunn of Georgia were clearly alarmed, with Nunn noting that:

> A close look at the Grenada operation can only lead to the conclusion that, despite the performance of the individual troops who fought bravely, the U.S. armed forces have serious problems conducting joint operations. We were lucky in Grenada; we may not be so fortunate next time.[28]

Incredibly, the experience in Grenada did little to spur the services into immediate action in fixing either SOF or joint warfighting doctrine. For the most part, the DoD remained locked into its focus on the conventional fight and procurement process, even as the Reagan administration began to shift some of its strategic focus towards smaller asymmetric conflicts. The DoD's apparent indifference garnered increasing ire from legislators, many of whom now felt that they were "preoccupied with chasing after resources" with "more time . . . spent preparing plans for the next budget than for the next war."[29] As the military bureaucracy continued to define SOF as "peripheral to the interests, missions, goals, and traditions that they view[ed] as essential,"[30] key legislators in Congress began to play an increasingly active role in addressing the "root causes" of the military's problems, that many felt were rooted in the current organizational structure of the Department of Defense. Emerging at the front of this effort to realize a joint-capable SOF component were Senators Cohen and Nunn and Representative Dan Daniel of Virginia. Ironically, as the Pentagon stubbornly resisted calls from these legislators for a "negotiated solution" to the unique issues facing SOF, each legislator shifted from supporting modest reform to the full reorganization of special operations forces under what would become USSOCOM.

Not surprisingly, DoD resistance to SOF reform or reorganization followed predictable reasoning, including an overwhelming focus on European-based conventional warfare, traditional distrust of an "elitist" organization, a belief that conventional "preparedness translates into a capacity to deal with "lesser included threats," and a "pervasive and persistent" fear that a re-energized SOF could again "drag us into another Vietnam."[31] Spotlighting the lingering fears of a conventional military, a Center for Defense information publication even cast doubt on modest ongoing SOF mission by suggesting that:

> Special Operations Forces training activities in Central America and elsewhere, like the Green Berets' training of South Vietnamese troops in 1959, could be the crucial first step on the path leading to direct U.S. involvement.[32]

From the DoD and the various services, bureaucratic intransigence towards Congress was "stiff and protracted, as is any defense of the status quo."[33] Legislative airing of SOF issues produced the:

> unintended effect of eliciting and illuminating some of the less attractive kinds of behavior associated with the military-careerism, parochialism, defensiveness—ills to which, in fact, any bureaucracy is heir, but which are seen as peculiarly egregious in this one.[34]

Tellingly, more modest internal initiatives fared no better. As early as 1982, Army Chief of Staff Edward Meyer attempted to bridge the divide between a standing mandate for conventional forces and the growing need for a standing SOF organization possessing substantial joint capabilities. His Strategic Services Command (STRATSERCOM), a modest proposal compared to the eventual radical reorganization that produced USSOCOM, was nevertheless soundly rejected by the Navy and the Air Force, who had little interest in even temporarily handing over their SEALS and aircraft to a standing joint organization.[35] As one DoD official caustically noted, the "biggest problem with STRATSERCOM was the Army invented it . . . so the Air Force went off and invented its own, which was called the Joint Operations Agency." Typical of the bureaucratic interservice turf wars, "after everybody played push-me-pull-you with it for a while, the whole business evaporated," returning the services to their status quo.[36] The entire process of substantial SOF reform had become so "agonizingly slow," that an exasperated Senator Barry Goldwater publically suggested that, "If we have to fight tomorrow, these problems will cause Americans to die unnecessarily [or] may cause us to lose the fight."[37] This failure to work with the Congress would prove a costly mistake for the Pentagon.

What legislative proponents increasingly noted as one of the most serious shortcomings facing special operations forces within the DoD structure was a "lack of effective advocacy" equivalent to that enjoyed by the respective services and even weapons platforms."[38] While SOF remained an issue discussed in professional circles and publications, support for change was ad hoc, and identifying a full-time advocate was challenging. Representative Daniel stepped in to fill the role of SOF advocate, which held profound consequences for the future of SOF. Aided by what a Congressional Research Service author understatedly referred to as "several able and persuasive staffers," Representative Daniel allied himself with Noel Koch, Principal Deputy Assistant Secretary of Defense for International Security Affairs, who was an ardent proponent of SOF inside the Pentagon and was less flatteringly described as the "Administration's 'rogue elephant'."[39] Together, their staffs set about to generate a controversy designed to place the SOF debate in the national, or at least legislative, spotlight. Using the pages of professional publications, they released a series of prepositioned opinion essays by a variety of authors advocating both reform and, less effectively, the status quo. The reform essays did not focus on a single argument, but were universally dedicated to fixing the SOF problem by restructuring SOF within the DoD and instituting a permanent advocate for SOF inside the DoD. Collectively, the articles succeeded in reinvigorating the debate over fixing SOF and would prove instrumental in shaping USSOCOM. Noel Koch's suggestions that "the short-term solutions are raising hell and politicking," seemed to be working.[40]

Between the release of the first advocacy article in August of 1985 and the introduction of a Senate amendment as part of the Goldwater-Nichols Defense Reorganization Bill, the DoD was in an entirely reactive posture regarding the debate over reinventing SOF. Whereas the narrative to this point placed most of the power in the hands of the DoD and conventionalists within the Armed Forces, the persistent inability and unwillingness of the Pentagon to seriously address the problem facing the doctrine, employment, and support of a consolidated SOF capability left key legislators little choice but to save the Pentagon from itself. Indeed, while Senators indicated that they "prefer it not be done by law," because "most of the organization could be done by the President [via the Secretary of Defense]," the Pentagon's track record since the end of Vietnam did not indicate it would willingly assist in instituting needed, or even directed, changes. Instead, Senator Cohen, who was now fully embroiled in the debate, indicated that the nation could "no longer temporize on the need to establish a clearer organizational focus for special operations and a clear line for their command and control."[41] Two competing bills, one from the House and one from the Senate, were introduced by Representative Daniel and Senator Cohen, respectively.

Much like the dueling advocacy essays from a year earlier, the dueling bills shifted the focus of the debate from the Pentagon to Congress. The two options, while significantly different, also concurred in the most important innovation in the command and control of SOF: the reorganization of all special operations forces under a joint entity other than their respective services. The status quo simply would no longer suffice.

Daniel's House bill placed the forces under a civilian-directed "National Special Operations Agency" (NSOA), while Cohen's Senate bill placed the forces in a "Special Operations Unified Command" under a four-star general. Having endured more reprogrammed funding on the part of the services, Daniel likewise stripped budget control from the parent services, while Cohen's bill left budgeting untouched. Finally, while Daniel's proposal had "self-contained" civilian oversight, Cohen's proposal dictated the formation of an "Assistant Secretary of Defense for Special Operations and Low Intensity Conflict" to provide needed advocacy.

In a telling exchange of who now controlled the debate over SOF, Chairman of the Joint Chiefs of Staff, Admiral William J. Crowe, presented both the House and the Senate with a hastily compiled alternative from the DoD. Arguing that "change was needed," but that "it should be a DoD responsibility to determine what form those changes should take and also implement them," Crowe suggested a toned-down alternative in a Special Operations Force Command headed by a three-star general who, as in Senator Cohen's plan, would report to the Secretary of Defense through the Joint Chiefs of Staff (JCS).[42] Claiming the DoD would have the JCS plan for SOF reorganization functional in just 6 months, he was met with a frosty reception in both chambers. Further "rigidly inflexible" efforts on the part of the DoD to promote the new JCS plan only hardened legislative resolve.[43]

Given a decade of friction involving the organization and funding of SOF, reconciling the approved House and Senate versions of the bill proved surprisingly easy. Agreeing to "focus on the objectives rather than the methodology," some of the shifts were made to fit more neatly within the larger Goldwater-Nichols Defense Reorganization Act.[44] The following provisions eventually became part of Public Law 99-661:

1. The creation of an Assistant Secretary of Defense for Special Operations and Low Intensity Conflict;

2. The formation of a unified combatant command for special operations forces (USSOCOM) and the placement of all active-duty and reserve special operations forces under USSOCOM;

3. The assignment of a four-star general or admiral to command USSOCOM;

4. The management of personnel and budgeting by the USSOCOM commander. Importantly, only the Secretary of Defense, in consultation with the SOF commander could place restrictions on the budgets approved by Congress;

5. The mandating that the SOF commander in unified combatant commands be a general or flag officer; and,

6. The establishment of a "Board for Low Intensity Conflict" within the National Security Council that the Senate recommended be headed by a new Deputy Assistant to the President for National Security Affairs for Low-intensity Conflict.[45]

Given the length of time that SOF received no billing, the changes encompassed in the Nunn-Cohen Amendment were sweeping and, for the Pentagon, severe. They combined nearly every legislatively-aggressive provision considered. While the House desire for a civilian-directed SOF agency within the DoD went unrealized, the establishment of a unified combatant command placed USSOCOM on par with just five combatant commands in the military. USSOCOM's power was further strengthened by the creation of Major Force Program (MFP) 11 to protect USSOCOM's funding from other resource-hungry services. Finally, in an acknowledgment of the increased potential for special operations forces, the Senate strongly advocated that the President restructure his National Security Council to include a Deputy Assistant for Low-intensity Conflict.[46] The changes could not have been a more stark contrast to SOF's precarious existence just a few years earlier.

After a decade of budgetary and structural omnipotence, the conventional forces within the DoD made a concerted, but ultimately futile, attempt to resist implementation of the provisions on the Nunn-Cohen Amendment. Most DoD foot dragging "generally reflected compliance with the letter rather than the spirit" of the legislative mandates, making for a halting and disjointed start for USSOCOM.[47] Because no USSOCOM commander had been immediately identified, the Navy attempted to retain command of its SEALs (an effort that ceased with the appointment of General James Lindsay at the first commander).[48] Although clearly intended to be headquartered in the Washington, DC, area, planners instead established USSOCOM at MacDill Air Force Base in Tampa, FL, and, when reluctantly manning it, filled it heavily with SOF-inexperienced personnel from the recently deactiviated U.S. Readiness Command.[49] More importantly, Secretary Weinberger steadfastly refused to fill the position of the new Secretary of Defense for Special Operations and Low Intensity Conflict. When under pressure from legislators to fill the obvious vacancy, he pointedly subordinated the position under another deputy and then filled it with Richard Armitage, a vocal opponent of restructuring and of special operations in general.[50] Even with the ostensible protection of MFP-11, various services continued to siphon early funds from SOF units that strayed away from the protective umbrella of USSOCOM.[51] While perhaps not always in "contravention of the law," as suggested by Senator Cohen, the "bureaucratic obstacles" erected by the Pentagon certainly slowed implementation.[52]

By this point, however, the balance of power rested with a Congress that was fully vested in the future of USSOCOM and committed to seeing "a proportionate response on the part of the military" to address "the increasing threat from the lower end of the conflict spectrum."[53] With the obvious realization that the DoD could not be cajoled into embracing the spirit of USSOCOM, Congress returned to its ability to forcibly force internal reorganizations on an unwilling DoD. Armed with the legislative pen, and "forced by bureaucratic resistance within the DoD to take very detailed legislative action," Congress produced Public Law 100-180 and Public Law 100-456, prescribing in great detail Congress' expectations from the Pentagon, to include an unprecedented amount of continuous congressional oversight.[54] The results, while not addressing all the issues facing USSOCOM and its legislated advocacy, continued to shape and reshape USSOCOM's mission, employment, and organization well into the 1990s.

While USSOCOM's continued presence by the mid-1990s indicated that it was firmly cemented into the security framework of the DoD, the road to achieving and sustaining a joint SOF capability was tortuous. Ironically, it was the very distaste of the DoD for SOF in the post-Vietnam era and the Pentagon's intransigence to reform that resulted in the bureaucratically radical formation of USSOCOM. USSOCOM's structure and orientation were in fact not driven by the unmistakable military necessity of low-intensity conflict, but by the very resistance of the DoD to acknowledge and address this necessity. Reformers from outside of the organization, recognizing both the Pentagon's uncorrected focus on the conventional possibilities of the Cold War and their dogmatic insistence on conventional strategy and procuring, at all costs, the weapons programs needed to win a conventional fight, rallied to the side of SOF employment. Offering SOF the effective advocacy previously denied them by the DoD, the fate of an entire unified combatant command would ultimately rest largely with several legislators.

ENDNOTES - CHAPTER 2

1. Congressional Research Service, "U.S. Special Operations Forces (SOF): Background and Issues for Congress," Washington DC: Government Printing Office, pp. 2-5.

2. General Accounting Office, Report to the Chairman, Committee on Armed Services, U.S. Senate: "Special Operations Command: Progress in Implementing Legislative Mandates," Washington, DC: U.S. Government Printing Office (hereafter GPO), September 1990, p. 8.

3. James Wooten for the Congressional Research Service, quoted in William Boykin, "Special Operations and Low-Intensity Conflict Legislation: Why Was It Passed and Have the Voids Been Filled?" Carlisle, PA: Individual Study Project, U.S. Army War College, 1991, p. 5.

4. Congressional Budget Office, "Outlays for Major Categories of Spending, 1968 to 2007, in Billions of Dollars" and "Discretionary Outlays, 1968 to 2007, in Billions of Dollars," *Revenues, Outlays, Deficits, Surpluses, and Debt Held by the Public, 1968 to 2007, in Billions of Dollars,* Washington DC: GPO, 2008, pp. 5, 7; Clinton Schemmer, "House Panel Formed to Oversee Special Operations Forces," *Armed Forces Journal International*, October 1984, p. 15.

5. William C. Ohl, "Fixing U.S. Special Operations: Rational Actors Not Allowed," Washington, DC: Individual Study Project, National War College, 1993, pp. 2-3.

6. Dan Daniels, "US Special Operations: The Case for a Sixth Service," *Armed Forces Journal International,* August, 1985, pp. 71-72.

7. Wooten, p. 5.

8. Henry A. Kissinger, *National Security Decision Memorandum 95: U.S. Strategy and Forces for NATO,* November 25, 1970, p. 1.

9. Robert T. Davis, "The Challenge of Adaptation: The U.S. Army in the Aftermath of Conflict, 1953-2000," Fort Leavenworth, KS: Combat Studies Institute Press, 2008, p. 52.

10. Deborah G. Meyer and Benjamin F. Schemmer, "An Exclusive Interview with Noel C. Koch, Principle deputy Secretary of Defense International Security Affairs," *Armed Forces Journal International*, March 1985, p. 52.

11. Charles R. Shrader, *History of Operations Research in the United States Army, Volume III: 1973-1995*, Washington, DC: GPO, 2009, p. 6.

12. Roy A. Grossnick, *United States Naval Aviation, 1910-1995*, Washington, DC: Naval Historical Center, p. 279.

13. Davis, p. 45.

14. *RESCUE MISSION REPORT* [Holloway Report], Washington, DC: U.S. Department of Defense, August 23, 1980, p. vi.

15. *Ibid.*, p. 15.

16. *Ibid.*, p. 16.

17. William G. Boykin, "The Origins of the United States Special Operations Command," Tampa, FL: Expanded Individual Study Project, U.S. Special Operations Command, n.d., p. 2.

18. On nuclear weapons, Michael Ganley and Benjamin F. Schemmer, "An Exclusive Interview with Senator Sam Nunn, Ranking Minority Member of the Armed Services Committee," *Armed Forces Journal International*, July 1986, p. 42. On threat in an age of deterrence, see Meyer and Schemmer, "An Exclusive Interview with Noel C. Koch," p. 52.

19. Secretary of Defense Caspar Weinberger, quoted in Daniels, p. 72.

20. Noel Koch, "Two Cases Against a Sixth Service," *Armed Forces Journal International*, October 1985, p. 108.

21. Boykin, "The Origins of the United States Special Operations Command," p. 8.

22. Meyer and Schemmer, "An Exclusive Interview with Noel C. Koch," p. 50.

23. Boykin, "The Origins of the United States Special Operations Command," p. 8.

24. William S. Cohen, "A Defense Special Operations Agency: A Fix for an SOF Capability That is Most Assuredly Broken," *Armed Forces Journal International (AFJI)*, January 1986, p. 39.

25. Meyer and Schemmer, "An Exclusive Interview with Noel C. Koch," p. 50.

26. *Ibid.*

27. Daniels, p. 70.

28. Senator Sam Nunn "Remarks" of October 2, 1985, as reprinted in "DOD Organization: An Historical Perspective," *Armed Forces Journal International*, October 1985, p. 15.

29. Senator Barry Goldwater "Remarks" of October 3, 1985, as reprinted in "Budget Dominance" The Quest for Dollars," *Armed Forces Journal International*, October 1985, p. 28.

30. On place of SOF within DoD, see Daniels, p. 72. On organizational structure and congressional response, see Nunn, "Remarks," p. 15.

31. John M. Collins, *Special Operations Forces: An Assessment*, Washington, DC: National Defense University Press, 1994, p. xxii; Cohen, p. 42; Daniels, p. 70.

32. Cohen, p. 44.

33. Koch, p. 103.

34. *Ibid.*, p. 102.

35. Boykin, "The Origins of the United States Special Operations Command," p. 4.

36. Meyer and Schemmer, "An Exclusive Interview with Noel C. Koch," p. 48.

37. Schemmer, p. 15; Senator Barry Goldwater "Remarks" of October 1, 1985, as reprinted in "Congressional Oversight of National Defense," *Armed Forces Journal International*, October 1985, p. 4.

38. Daniels, p. 72; Ganley and Schemmer, p. 42.

39. Collins, p. 22; Deborah G. Meyer and Benjamin F. Schemmer, "Congressional Pressure May Force Far More Dollars for Special Ops," *Armed Forces Journal International*, April 1986, p. 20.

40. Koch, "Two Cases Against a Sixth Service," p. 108.

41. Cohen, "A Defense Special Operations Agency," p. 45.

42. Boykin, "The Origins of the United States Special Operations Command," p. 11.

43. Susan Marquis, *Unconventional Warfare: Rebuilding U.S. Special Operations Forces*, Washington, DC: Brookings Institution Press, 2007, p. 140.

44. Boykin, "The Origins of the United States Special Operations Command," p. 14.

45. "Special Operations Matters," *Department of Defense Authorization Act, 1987*, Public Law 99-661, 99th Cong., 2nd sess., November 14, 1986, Sec. 1311.

46. *Ibid.*

47. Collins, p. 12.

48. *Ibid.*, p. 13.

49. Walter Fischer, "Why More Special Forces are Needed for Low Intensity War," *Heritage Foundation Report*, November 12, 1987, p. 8.

50. Collins, p. 11.

51. FY89 *Department of Defense Authorization Act*, Public Law 100-456, 100th Cong. 2nd sess., September 29, 1988.

52. David Ottaway, "Delay on Guerilla Command Irks Hill; Weinberger Wants 12th Assistant Secretary's Post Created," *The*

Washington Post, March 10, 1987; General Accounting Office, *Report to the Chairman, Committee on Armed Services, U.S. Senate,* "Special Operations Command: Progress in Implementing Legislative Mandates," Washington, DC: GPO, September 1990, pp. 13-14.

53. Cohen, p. 45.

54. Collins, p. 11.

CHAPTER 3

SAILING ON STORMY SEAS: U.S. JOINT FORCES COMMAND AND REORGANIZATION IN THE POST-COLD WAR WORLD

Seanegan P. Sculley

INTRODUCTION

From 1993 to 2002, United States Atlantic Command (USACOM), formerly Atlantic Command (LANTCOM), implemented a reorganization initially meant to foster a new ethos of joint operations throughout the Armed Services community. USACOM quickly implemented significant changes during a period of heightened tension within one of its Areas of Responsibility (AOR), and did so efficiently and effectively. During the next 8 years, however, regions were reallocated to other commands, and USACOM became increasingly functional in focus. In the wake of a new massive reformation throughout the Defense community in 2002, USACOM lost all of its operational capacity and became the purely functional command U.S. Joint Forces Command (USJFCOM). For the next decade, the command positioned itself as the prominent laboratory for transformation concepts. In 2002, USJFCOM ran the most expensive and extensive exercise in U.S. history to test those theories, MILLENNIUM CHALLENGE 2002, but with mixed results. Now, in 2011, USJFCOM is being de-commissioned. Arguments about the reasons and effects of this reorganization are polarizing. Some argue that in an era of limited resources, commands without an operational focus are superfluous, while others suggest that USJFCOM has successfully fulfilled its functional purpose. The most compelling argument may lie between these two.

In the early fall of 1994, a large American force was gathering against a small Caribbean nation. Paratroopers from the 82nd Airborne Division mustered on the "Green Ramp" at Pope Air Force Base, Fayetteville, NC, to be issued their ammunition and board C-130 cargo aircraft. Soldiers from the 10th Mountain Division were already aboard the USS *Dwight D. Eisenhower* off the coast. Naval aviation, Air Force fighters, and Army helicopters stood by, while Marines prepared for a possible amphibious landing. Operation UPHOLD DEMOCRACY was about to commence.

Simultaneous to the military prepositioning of forces, a small group of diplomats was meeting with Haitian leaders. In January 1991, Jean Bertrand Aristide was instated as President of Haiti. His tenure was cut short, however, by a military junta that overthrew Aristide's government and controlled Haiti for more than 3 years. In an attempt to reinstate President Aristide, the U.S. Government under President William Clinton imposed tough economic sanctions, which created a flood of refugees from Haiti and, it was hoped, brought the junta to a more congenial position with regards to American demands.

Former U.S. President Jimmy Carter and former Chairman of the Joint Chiefs of Staff Colin Powell led the diplomatic contingent to Haiti. They informed their hosts that American paratroopers were in the air and the full military might of the United States would be brought to bear against the Haitian government if Aristide was not allowed to return to the island and peacefully assume his role as President. Faced with overwhelming odds and certainly reflecting on American success in Panama (and more recently in Iraq), the military leaders of the junta capitulated and, for the most part, violence was avoided.

Operation UPHOLD DEMOCRACY was the first military operation for a new unified command reorganized just months earlier under the Goldwater-Nichols Department of Defense Reorganization Act of 1986. U.S. Atlantic Command (USACOM) was created to do more than act as a headquarters for commands from the various services, and its responsibilities were larger than the regions over which it had control. Formed under the supervision of Powell in the early-1990s, USACOM was the new unified command tasked as the advocate for joint training, joint provision, and joint integration for all continental U.S. (CONUS)-based forces. It was organized to provide trained joint forces from the United States to other combatant commands around the world if mission requirements exceeded the internal forces of the supported command; additionally, it was responsible for the defense of North America, the Atlantic Ocean, and the Caribbean Sea.

Rather than form a new unified command to accomplish this goal, Powell decided to give the mission to Atlantic Command (LANTCOM) and allow its commander in chief (CINC), Admiral Paul D. Miller, to reorganize to meet the new mission requirements. This feat was not accomplished in a vacuum. LANTCOM had an AOR that spanned the Atlantic Ocean and Caribbean Sea. Missions were ongoing in the Caribbean due to the escalating tensions with Haiti and responsibilities still existed from the Cold War towards the Azores and Iceland. Yet in the spring of 1993, LANTCOM staff developed a sound implementation plan that, within months, was accepted by the Secretary of Defense and the President, and allowed for the success just 1 year later of Operation UPHOLD DEMOCRACY.

The following years would continue to change USACOM's structure and mission. Gradually, AORs were given to other commands. Initially, the command surrendered the Caribbean to U.S. Southern Command (SOUTHCOM); this move was followed by the loss of the Atlantic to U.S. European Command (EUCOM); finally, in 2002, USJFCOM was divested of its role as defender of CONUS with the formation of U.S. Northern Command (NORTHCOM) after the terrorist attacks of September 11, 2001 (9/11). At the same time, new functional organizations were placed under its control, refining its focus as the joint advocate for all U.S. forces.

In October 2002, after 9/11 and the rapid escalation of the Global War on Terror (GWOT), USACOM was renamed U.S. Joint Forces Command (USJFCOM) and divested of all of its operational requirements. The command accepted its role as a purely functional command, focused on developing concepts derived from the Pentagon, Washington, DC-directed programs of Transformation and the Revolution in Military Affairs (RMA). This focus was best exemplified in the largest joint exercise in history, MILLENNIUM CHALLENGE 2002 that produced mixed results and one very public controversy. Additionally, many of those opposed to the concept under which USJFCOM was formed remained hostile, and some who were responsible for USJFCOM's creation ultimately joined its enemies' camp. While the reorganization of USACOM in 1993 was a success, the eventual reality in 2002 of USJFCOM as a purely functional command focused on joint operation advocacy and transformation limited its perceived relevance in the GWOT and allowed those in opposition to the concept to move for its disestablishment.

THE NAVY'S BACKYARD: ATLANTIC COMMAND AND THE COLD WAR, 1947-90

Atlantic Command (LANTCOM) was originally created under the Unified Command Plan of 1947.[1] By the early-1950s, LANTCOM evolved into a largely naval command responsible for the defense of the North Atlantic against a growing Soviet threat. Sub-unified commands were created in the Azores and then Iceland, to provide air and naval bases from which to launch forces and keep sea lanes clear. LANTCOM's command structure revealed this focus clearly: the CINC for LANTCOM commanded the main component command, the Atlantic Fleet (LANTFLT) and was the Supreme Allied Commander for the Atlantic (SACLANT) under NATO. As Soviet submarine and surface fleet capabilities grew, LANTCOM's importance increased and it played a key role in the Cold War.[2]

While it played a chess match in the North Atlantic, LANTCOM gradually gained responsibilities closer to home. In the Caribbean, Communism became a constant threat to the small nations of the Antilles and Central America. Over time, LANTCOM took over defense of the Panama Canal, then the waters of the Caribbean and, eventually, the defense of the Caribbean islands. This expansion of responsibilities occurred in the 1960s, a time of growing tensions between Cuba and the United States. LANTCOM was integral to the naval blockade of Cuba in 1962 during the Cuban Missile Crisis and took control of Joint Task Force (JTF) 120 and 122 during the 1965 invasion of the Dominican Republic.[3]

These operations continued to illustrate the difficulties inherent in joint operations. Starting in 1963, LANTCOM began hosting large joint training exercises between different U.S. services and often included allied forces as well. Originally named Quick Kick, these biannual events evolved into a large-scale field exercise off the East Coast of the United States called Solid Shield. Conducted in the 1980s, Solid Shield involved between 40,000 and 50,000 personnel from the Army, Marine Corps, and Navy, with a focus on the rapid deployment of joint forces within LANTCOM's AOR. This focus on joint operations led to the selection of LANTCOM to lead the mission into Grenada in 1983.[4]

Similar to the subsequent intervention in Haiti, Operation URGENT FURY was a large-scale joint mission that attempted to include every service in the U.S. military, including Special Operations forces from both the Army and Navy. While the mission was a success, many in the Armed Services and in Congress were critical of the logistical planning and execution, believing either that only one service should have been involved or that American military joint capabilities needed reform. While historians argue over the initial catalyst for change, the complications in the execution of the invasion in Grenada undoubtedly helped secure the passage of the Goldwater-Nichols Act.[5]

The intent of the Goldwater-Nichols Act of 1986 was to encourage joint cooperation between the Armed Services and streamline joint operations in the future, both of which had been attempted unsuccessfully in the past. Both Strategic and Tactical Interdiction Keeper of Eagle Command (STRIKECOM) and U.S. Readiness Command (USREDCOM) in the 1960s and 1970s had been created to support joint operations, but neither command possessed the structure to support the mission and overcome interservice rivalries.[6] Under the auspices of Goldwater-Nichols, the Chairman of the Joint Chiefs of Staff (CJCS) had new authority. Not only would he have a direct line to the Office of the President as chief military advisor, the CJCS was also in charge of joint operation doctrine and, at the request of the President, could be used as a conduit of communication between the CINC and his unified commanders. Furthermore, the law defined the roles of the unified commanders and encouraged the Service Chiefs to place their best and most promising officers in joint assignments.[7]

These new legislative guidelines were intended to foster a spirit of "jointness" between the services, to encourage effective and efficient military operations that would combine air, land, and sea forces to defeat enemy nation-states in a large-scale, conventional war. The debate over the best way to accomplish this goal was heated and split the services while, in many instances, uniting politicians across party lines. The U.S. Army and Air Force were largely in favor of the new act as it was drafted, but there were accusations made by staffers for the Senate Armed Services Committee that the Department of the Navy actively opposed the bill.[8]

Simultaneous to the debate in Congress over the passing of the new law, the Chief of Naval Operations requested in 1985 that the staffs and commanders for LANTCOM and LANTFLT be separated to allow the commander-in-chief LANTCOM (CINCLANT) to concentrate on the increasingly complicated role he was playing. LANTFLT would remain a component command under LANTCOM but would not continue to fulfill its de facto role as the unified staff for all of LANTCOM.[9] In 1989, the role as commander of Atlantic Command became more difficult with the end of the Cold War.

A FIGHT FOR RELEVANCE: USACOM FINDS ITS PLACE IN THE GOLDWATER-NICHOLS REORGANIZATION AND THE ENDING OF THE COLD WAR

In August 1991, as the new Chairman of the Joint Chiefs of Staff, General Powell held his annual CINC Conference. At that meeting, he proposed a vigorous implementation of the reorganization within the Armed Services he believed was intended in the Goldwater-Nichols Act. Despite military successes in Panama and Iraq in 1989 and 1991, Powell knew budgetary restrictions were on the horizon and much of the U.S. force would be returned to CONUS bases. This would further complicate projection of joint forces globally and would require a focused approach to the problem. To solve it, he proposed a new unified command dedicated to joint force advocacy, and he had a command and commander in mind.[10]

Admiral Paul D. Miller was a native Virginian and the CINC for the Atlantic Fleet in 1991. Powell described him as "a very aggressive and brilliant commander" and a true believer in the concept of joint operations.[11] Though Powell's suggestions to reduce the number of unified commands from 10 to six and create a new unified command in charge of all the Americas was rejected by the CINCs at the annual CINC Conference in 1991, that next summer he placed Miller (then CINCLANT) and his staff in charge of creating a plan to implement the necessary efficiencies and provide a joint force advocate to change the culture of the American military.[12]

During this time, other CINCs and members of Congress, including Chairman of the Armed Services Committee, Senator Sam Nunn (D-GA), voiced their agreement that reorganization of the Armed Services was necessary to meet growing international requirements in the face of shrinking resources. At the CINC Conference in August 1992, Powell submitted his idea to place all CONUS-based forces under a single unified command, and Miller seconded the idea, suggesting LANTCOM as the unified command. By the end of the conference, the CINCs had agreed to research three possibilities: 1) keep current organization of forces; 2) give the new Air Combat Command (ACC) specified command status (allowing the ACC to command and control all Air Forces in the United States); or 3) reorganize LANTCOM to command all CONUS-based forces as the joint force integrator.[13]

Miller placed his special assistant, Navy Captain William D. Center, with the Joint Staff Strategic Plans and Policy directorate (J5) to research and write a concept paper discussing a unified command focused on joint force advocacy. The plan that was eventually devised would increase the scope of LANTCOM drastically by placing all CONUS-based forces under its command to facilitate its new role as joint force **provider** and **trainer** to other unified commands around the world.[14] Powell agreed with the concept. He viewed LANTCOM as the best choice for the assignment. It had a history of joint cooperation over the past 3 decades in the Caribbean. It had a reduced mission requirement as the defender of the Atlantic following the fall of the Soviet Union. The new commander, Miller, was an adherent of Powell's joint concept and willing to do the work needed to regain relevance for his command.[15]

In a March 1993 memorandum to Secretary of Defense Les Aspin, Powell argued for the reorganization of LANTCOM as the joint force integrator of America's CONUS-based force. On March 29, 1993, Aspin wrote a letter to Nunn, agreeing with Powell's argument that such a reorganization would create efficiencies and unity of effort in the realm of joint operations.[16] At the same time, Miller's staff began planning for implementation, knowing that final approval would require an implementation plan acceptable to the majority of the CINCs and Service Chiefs.

The LANTCOM staff created a LANTCOM Implementation Working Group (LIWG), within the command and outside of it with other CINCs and their staffs to quickly write a draft plan that incorporated input from the entire LANTCOM staff and from the component commands. To do so, the organization of the LIWG was based on a cross-functional model. Seven functions of the new organization were identified, and the directorates of LANTCOM most closely associated with those functions were assigned leadership of the development teams. Each team had a general officer/flag officer as a team leader and a colonel as the team director. The teams created a list of "kick start" issues they considered of primary importance to allow for quick action once implementation started because they feared a period of inaction often associated with the formation of new groups.[17]

While the plan was being drafted, LANTCOM held periodic general officer/flag officer review group meetings with the CINCs and their staffs from the proposed component commands. As their concerns and suggestions were included, the draft plan was sent to the staffs of the CINCs for the other unified commands and then to the CINCs themselves. Miller knew these commanders would initially have serious doubts about the proposed plan and hoped to short-circuit such objections by giving each unified command CINC the opportunity to voice concern during the planning process. All of this was designed to allow for the greatest amount of participation and inclusion while accepting a very short planning cycle.[18]

At the same time, in the period of November 1991 to June 1993, Haiti was in the throes of a military coup. Following the imposition of economic sanctions by the United States, Haitian refugees attempted to flee to America in makeshift boats. An injunction blocked any attempt to return these people to their home island and a refugee camp was established at Guantanamo Bay (GTMO), Cuba. LANTCOM was in charge of the joint task force running the operation (JTF-GTMO), and over the next 2 years, the population in the camp fluctuated from a few hundred to over 15,000 men, women, and children.[19]

As the CINCs took the time to examine the proposed implementation, several observations were forwarded and considered. U.S. Transportation Command (TRANSCOM) wanted to be included in all joint planning and was unwilling to give up command of Air Force lift assets. U.S. Central Command (CENTCOM) was concerned because U.S. Pacific Command (PACOM) was not going to be forced to relinquish its control of CONUS-based forces on the West Coast. CENTCOM also feared any joint force packages it received from LANTCOM would not include the best forces for the mission. Not only would LANTCOM gain command of the four main service-related force commands in the United States (U.S. Army Forces Command (FORSCOM), LANTFLT, ACC, and Marine Forces Atlantic (MARSFORLANT), it would also be responsible for the continental defense of the United States. In this role, U.S. Space Command (SPACECOM) insisted that North American Aerospace Defense Command (NORAD) be included in the planning. Special Operations Command (SOCOM) refused to give LANTCOM authority to include Special Operations Forces (SOF) in their joint force packages, and many wanted LANTCOM's AOR greatly reduced to give its staff the ability to focus on the complexities of its new mission.[20] In short, Miller's staff was threatening to break rice bowls, and the CINCs were not willing to sit quietly while it happened.

The LIWG took into account the concerns of the other unified commands and presented the draft Implementation Plan to the CINCs again in May 1993. The concerns addressed in the new draft were consolidated into three main objections, and the plan was forwarded to the Joint Staff and the Service Chiefs. These objections focused on the fact that USACOM (the new acronym for LANTCOM under the Implementation Plan) would not command all forces in the U.S. (PACOM would retain command of forces on the West Coast), that few commanders understood the new concept of lead operational authority created by the LIWG to explain how component commands would run routine missions, and that there was fear the concept of adaptive joint force packages would not be flexible enough to meet the needs of supported commanders.[21] The Service Chiefs from the Army and the Air Force agreed with and supported the plan while Chief of Naval Operations Admiral Frank Kelso did not. He, along with the Commandant of the Marine Corps, believed this reorganization would drain money from the Department of the Navy to the benefit of the other services and not add a benefit to naval forces. The Marine Corps contended it already conducted joint operations and did not need a separate unified command to act as trainer, provider, or integrator.[22] Despite these objections, the plan was approved and signed into the Unified Command Plan on September 24, 1993, by President Clinton. The rebirth of LANTCOM as USACOM was celebrated in a ceremony on October 1, 1993.

The fact that a plan of this magnitude could be conceived, created, and approved in a year is astounding. The cultural and organizational ramifications of the new USACOM as joint trainer and provider were profound. The use of a functional organization model and an adherence to inclusion of the entire unified command community were responsible for the success of this achievement. For the next 10 years, that implementation plan would be the basis for a successful command, one that expanded quite quickly, but one that ran into some unforeseen obstacles early on as well.

STAKING ITS CLAIM ON AN UNCERTAIN FUTURE: TRANSITIONING DURING THE GLOBAL WAR ON TERROR

On the day of Powell's retirement, October 1, 1993, USACOM—the giant—was born. It now not only had responsibility for most of the Western Hemisphere, it also commanded the majority of American military forces. At the same time, American diplomats believed they had reached a peaceful solution to the problem in

Haiti. The junta in charge of the island agreed to allow Aristide to return with American forces to aid in a political transition. But when the USS *Harlan County*, carrying the initial contingent of U.S. forces, was turned back in the fall of 1993, the Joint Staff began planning for a force-based solution.[23]

Two operational plans (OPLANS) were devised by USACOM into 1994. One was for a forced entry into Haiti (JTF-180) and the other was in case a peaceful solution was reached (JTF-190). With both JTFs positioned in September of 1994, the stage was set for the successful diplomatic mission of President Carter and General Powell. Since both JTFs were ready, the actual operation that reinstated Aristide to power was a combination of forces from both.[24] Despite the success of Operation UPHOLD DEMOCRACY from the standpoint of joint operations, the Secretary of the Navy and the Chief of Naval Operations were able to convince the Secretary of Defense to alter the Unified Campaign Plan (UCP) in 1995 and remove the Caribbean from USACOM's AOR; it was given to SOUTHCOM to streamline drug interdiction operations and became effective in January 1997.[25]

This shift was, arguably, the most important when looking toward the future for this newly reorganized command. From its inception, USACOM failed to gain the support of its former master, the U.S. Navy. The only service to maintain its objections against the concept of a unified command focused on joint operation advocacy was also the service charged with financing USACOM. Throughout its existence, the command remained woefully short on manpower. In an interview in 2003, the man responsible for standing up the Joint Warfighting Center, Eugene Newman, stated his two biggest concerns in implementing joint training were a lack of forecasted funds and perpetual under manning.[26] Given its limited resources, it may not be surprising that USACOM lost its control of the Caribbean. Yet the precedent that was set, and a growing institutional belief that the future of the command lay in its functional role may have begun a 5-year slide toward a complete loss of operational responsibilities.

In other ways, though, the command was still expanding. It created a new facility to aid its reorganized Joint Training Directorate (J7). The Joint Training, Analysis, and Simulations Center (JTASC) focused on training commanders and their staffs in joint missions at the operational level and worked primarily through computer simulations. Staffing remained a constant problem in USACOM and to combat this, U.S. Air Force Major General Michael Short (J7) devised a method of incorporating reservists during peak training times in the Joint Reserve Unit (JRU). Then in 1995, USACOM created the Joint Processing and Onward Movement Center (JPOM) at Fort Benning, GA, to deploy joint forces from CONUS to the Balkans.[27]

Following the success of these joint ventures, in 1998, five Chairman Controlled Activities all focused on joint integration and training became part of the USACOM organization.[28] Yet one of the key concepts for USACOM as envisioned by both Powell and Miller failed to materialize. Adaptive Joint Force Packaging (AJFP) was not accepted by the combatant commanders overseas. The idea of training habitual joint units in different configurations to meet possible contingencies was considered too rigid by the other unified commanders. They wanted the power to decide what force structures they needed for their mission sets. They were not willing to trust USJFCOM to predict all possible contingencies and build packages to meet those outcomes. Instead, an ad hoc system was devised that involved the Joint Staff, the supported command, and the supporting command and allowed other unified commands, like PACOM, to also fulfill the role of joint force provider and give the supported command greater latitude in prescribing manning requirements.[29]

Still, by 1999 USACOM was fully engaged in its functional role as joint force trainer, provider, and integrator. It gained another three Chairman Controlled Activities focused on joint functions. At that point, to recognize this fact and to highlight the future of the organization, USACOM was renamed in the UCP 1999 as U.S. Joint Forces Command (USJFCOM). Additionally, it still had substantial regional responsibilities, not the least of which was the defense of CONUS. So when the terrorist attacks on September 11, 2001 occurred, USJFCOM was ready to provide joint task forces to aid New York City. Though these forces were not required, USJFCOM did work to stand up a JTF Headquarters for Homeland Security.[30] Shortly thereafter, however, a new reorganization of the American defense community would come and leave USJFCOM with no operational responsibilities whatsoever.

In the new UCP 2002, the Atlantic region was taken from USJFCOM and given to U.S. European Command (EUCOM) and defense of CONUS was given to the newly formed U.S. Northern Command (NORTHCOM), Thus, USJFCOM was left as a solely functional command.[31] In effect, the attacks on 9/11 led to a reduction in USJFCOM capabilities first demanded by the other unified commands during implementation planning in 1993. In the space of less than 10 years, following an unprecedented attack on U.S. soil, the largest, and possibly most powerful, unified command had been relegated to training, integration, and experimentation.

SAILING INTO PORT OR SINKING TO THE DEPTHS? THOUGHTS ON THE DISESTABLISHMENT OF USJFCOM

USJFCOM was the one unified command to be cut in another era of budgetary cuts that came in the early-2010s. This decision came less than 20 years after it was designated the premier command for a transformational military of the future. Yet command historian William McClintock has some thoughts on the structural weaknesses inherent in USACOM that may have led to this situation. First, USACOM, though a large unified command, remained under the budget of the Navy, which was the main opponent to the idea of a joint forces command from the outset. The Navy's budget was shrinking in 1993, and the Chief of Naval Operations had already given his opinion that the new USACOM brought no value-added to his service. With this in mind, it is not surprising USACOM also habitually encountered manning problems. It was always understaffed for its mission. Second, USACOM did not control the budgets of its component commands. Without the power of the purse strings, it is difficult to imagine directing FORSCOM or MARSFORLANT to do something contrary to the wishes of its service chief. Finally, while the original concept placed all CONUS-based forces under USACOM, the reality was this never happened. Instead, these forces were split between east and west coasts. Without the full complement of forces, it was difficult for USACOM to maintain a position as the joint force provider.[32]

Nevertheless, USACOM was successful through the 1990s and into the new millennium. What happened after 9/11? An interview with the director of the Joint Warfare Center in 2003, Dr. Eugene Newman, is enlightening. He claimed the quality of officers arriving at the center for training had changed for the better. In 1993, officers often arrived without computer skills, without joint experience, and without an understanding of why joint training was necessary. Ten years later, he noted none of that was true. It was now foreign to American officers to think outside of joint operations; if the mission did not involve joint forces, it did not pass muster.[33] Perhaps the time for a dedicated force to serve as the joint force trainer and integrator has passed precisely because USJFCOM succeeded in elevating "jointness" as a mandatory consideration in military operations. Secretary of Defense Robert Gates said as much in February 2011, when he claimed that joint cooperation was now fully embedded in American military ethos and so the need for a separate command dedicated to joint advocacy had passed.[34]

The manner with which the Office of the Secretary of Defense (OSD) approached the dismantling of JFCOM belies a successful completion of mission. Senators and Congressmen from Virginia, concerned about the 5,000 jobs at stake with the dissolution of the command, claim Gates did not provide adequate analysis or transparency for his decision. They threatened to subpoena him to testify before the House Armed Services Committee to justify his actions. Congressional outrage lessened after an agreement was made to keep approximately half the jobs threatened, transferring select organizations within JFCOM to other commands identified as fulfilling still relevant functions for the military.[35]

Still, the decision to disestablish USJFCOM was made quickly and did not appear to include the input of the Armed Services or Congress. The idea was first presented at the Public Session Quarterly Meeting for the Defense Business Board (DBB). The DBB was established by the Secretary of Defense (SECDEF) to find inefficiencies that would streamline the defense budget. Retired Marine Corps Major General Arnold Punaro led the committee and delivered the first blow to JFCOM. In his review, Punaro claimed JFCOM was riddled with redundant commands and should be eliminated to save the Department of Defense (DoD) money. He announced this recommendation on July 22, 2010.[36] Within a month, in August 2010, Gates appeared on various news outlets and stated he would recommend to the President that JFCOM should be dissolved. By January 6, 2011, the new UCP was signed by President Barak Obama, and the decision was finalized.

It is interesting that Punaro would be the person leading the charge to sink JFCOM. He spent 24 years as a congressional staffer to Senator Nunn (D) where he was neck-deep in the fight to gain the votes necessary within the Armed Services Committee to approve the Goldwater-Nichols Act. His actions earned him condemnation from members of his own service in the Pentagon for supporting the bill.[37] From 1994 to 1997, Punaro served on the General Officer's Steering Committee for the CINC, USACOM. It would appear he was one of those in Washington responsible for the creation of JFCOM, a loyal proponent for the joint advocacy Powell envisioned in the early-1990s. Thirteen years later, he had completely reversed his position with regards to the command.

A possible explanation could lie with the largest exercise ever executed in American history—MILLE-

NIUM CHALLENGE 2002—a massive undertaking meant to test joint concepts devised by JFCOM. It was both an exercise and an experiment, incorporating computer simulations and field exercises to create a joint operation focused on the situation in the Middle East to test initiatives based upon Revolution in Military Affairs tenets, particularly the concept of Effects-Based Operations.[38] MILLENNIUM CHALLENGE 2002 took 2 years to plan, involved 13,500 personnel on 26 different installations and cost the taxpayer $250 million.[39]

While a spokesman from USJFCOM claimed the experiment was a success, and the concepts tested were sufficiently validated for acceptance as doctrine by the DoD, a senior-ranking member of the exercise publicly denounced this assertion. Retired Marine Lieutenant General Paul Van Riper (U.S. Marine Corps [USMC]) was hired as a contractor to play the part of the opposing force (OPFOR) commander. For the first 3 days of the 3-week-long exercise, Van Riper was allowed to determine all of his actions relatively unfettered. He exploited over-reliance on technology by the U.S. forces to sink much of the naval force arrayed against him and caused those running the experiment to re-set the simulation. Van Riper stated in news articles in the months following the exercise that he decided to quit as the OPFOR commander when it became clear he would be required to follow a very restrictive script that would allow the U.S. forces to win the game. He claimed the concepts failed testing or were not properly vetted, and he would not allow his name to be attached to their validation and use by the armed forces. He made further assertions that he had witnessed the same absence of intellectual integrity in a previous exercise, Unified Vision 2001, an exercise also devised by USJFCOM.[40]

While there is no proof that Van Riper spoke with Punaro privately about his concerns, it is probable that they have known each other since the Vietnam War.[41] Regardless, it appears USJFCOM had suffered a serious setback with the public and costly embarrassment following MILLENNIUM CHALLENGE 2002 that could have changed Punaro's position regarding the relevance of the command. Furthermore, throughout the first decade of the 21st century, the command wedded itself to Secretary of Defense Donald Rumsfeld's vision of Transformation, a model based on tenets of a so-called Revolution in Military Affairs. These tenets have since been largely discredited in many circles of the military and do not seem to mesh well with Secretary Gates's vision for the future military as enumerated in his recent comments at the various service academies.[42] Combined with a prolonged war, a growing debate over the federal deficit, and an overall acceptance by the services of a joint ethos, USJFCOM's adherence to Transformation concepts provided its critics with an opportunity to torpedo the command. Whether through success or irrelevance, the niche USJFCOM created for itself has disappeared.

ENDNOTES - CHAPTER 3

1. Leo P. Hirrel and William R. McClintock, *United States Joint Forces Command Sixtieth Anniversary, 1947-2007*, Norfolk, VA: Office of Command Historian Joint Forces Command, 2007, p. 3.

2. *Ibid.*, pp. 7-9.

3. *Ibid.*, pp. 11-18.

4. *Ibid.*, pp. 19-26.

5. *Ibid.*, p. 28.

6. William R. McClintock, *Establishment of United States Atlantic Command, 1 October 1993*, Norfolk, VA: Headquarters, Commander in Chief U.S. Atlantic Command, 1996, p. 10.

7. Goldwater-Nichols Department of Defense Reorganization Act of 1986, Section 201, § 153, Washington, DC: Department of Defense.

8. James L. Locher III, "Goldwater-Nichols: Fighting the Decisive Battle," *Joint Forces Quarterly*, Summer 2002, p. 42.

9. Hirrel and McClintock, pp. 28-29.

10. Colin Powell, "The Chairman as Principal Military Advisor: An Interview with General Colin Powell (United States Army, Retired), former Chairman, Joint Chiefs of Staff," *Joint Forces Quarterly*, 1996.

11. *Ibid.*

12. McClintock, pp. 7-9.

13. *Ibid.*, pp. 9-10.

14. *Ibid.*, pp. 9-11.

15. Powell.

16. Les Aspin, Letter to Senator Sam Nunn, March 29, 1993.

17. McClintock, pp. 22-23.

18. *Ibid.*, pp. 22-24.

19. Hirrel and McClintock, pp. 31-32.

20. McClintock, pp. 27-28.

21. *Ibid.*, pp. 47-50.

22. *Ibid.*, pp. 50-51.

23. Hirrel and McClintock, p. 32.

24. *Ibid.*, pp. 33-34.

25. *Ibid.*, pp. 37-39.

26. Leo Hirrel, Interview with Mr. Eugene Newman, Joint Warfare Center, King George, VA, December 4, 2003.

27. Hirrel and McClintock, pp. 43-47.

28. *Ibid.*, p. 48.

29. *Ibid.*, p. 46.

30. *Ibid.*, pp. 51, 60.

31. *Ibid.*, p. 60.

32. McClintock, pp. 67-71.

33. Hirrel.

34. Tracy Agnew, "Suffolk's JFCOM Workforce to Shrink," *Suffolk News-Herald*, February 10, 2011.

35. Robert McCabe, "General: JFCOM Downsizing to be Complete by March 2012," *Virginian-Pilot*, February 10, 2011.

36. Arnold Punaro, "Reducing Overhead and Improving DoD's Business Operations," *Public Session Defense Business Board*, Quarterly Meeting, July 22, 2010.

37. Locher, p. 44.

38. Hirrel and McClintock, pp. 58-59.

39. Sam D. Naylor, "War Games Rigged? General says MILLENNIUM Challenge 02 'was almost entirely scripted'," *Army Times*, August 16, 2002.

40. *Ibid*. Popular author Malcom Gladwell also discusses Van Riper and the failures of MILLENIUM CHALLENGE in his book, *Blink: The Power of Thinking without Thinking*, New York: Little, Brown, 2005, pp. 99-146.

41. Congressional Record, Vol. 149, Issue 146, October 17, 2003, S12833. Punaro's company commander in Vietnam, Colonel (Ret.) Jim Van Riper, is Lieutenant General Paul Van Riper's twin brother and was present at Punaro's retirement ceremony in September 2003.

42. Robert Gates, "Speech Delivered to West Point Cadets," West Point, NY, February 25, 2011.

CHAPTER 4

ORGANIZATIONAL INSECURITY AND THE CREATION OF U.S. CENTRAL COMMAND

James C. Harbridge

INTRODUCTION

U.S. Central Command (USCENTCOM) has its origins in the turbulent social and political events of the 1970s in the Middle East. These events fundamentally altered the base upon which American foreign and military policy rested. In response, the Rapid Deployment Joint Task Force (RDJTF) was created. The creation of the RDJTF caused much interservice and intraservice conflict, which hampered its effectiveness. In the early-1980s, RDJTF was reconstituted as U.S. Central Command (USCENTCOM), a permanent unified command with regional responsibility. The basic culture and organizational structure, however, did not change, and this stability presented significant challenges to an organization that consistently felt the need to prove itself. Thus, the culture of insecurity and justification that existed within RDJTF persisted in USCENTCOM. Indeed, the culture persists to this day and affects operations and command culture.

At midnight on January 1, 1983, U.S. Army Lieutenant General Robert C. Kingston assumed command of USCENTCOM, located at MacDill Air Force Base in Tampa, FL. The unusual timing resulted from the fact that this was not a change of command; it was, rather, the assumption of command of an "entirely new" unified command. In reality, USCENTCOM was organizationally the same on that New Year's morning as it had been when it was the RDJTF on New Year's Eve. This fact was not lost on any of the Commanders in Chief (CINCs) of the other unified commands—of U.S. European Command (USEUCOM), U.S. Atlantic Command (USLANTCOM), and U.S. Pacific Command (USPACOM)—nor was it lost on the Joint Chiefs of Staff (JCS) and its staff, or indeed, the commander and staff of USCENTCOM itself.

The creation of USCENTCOM as one of five geographically oriented major commands marked the culmination of an organized 3-year plan within the Department of Defense (DoD). This plan aimed at addressing the unpreparedness of the United States to deal with the geopolitical realities of Southwest Asia that became apparent in the 1970s. Ultimately, rivalries both among the four services and within individual services played a major role in the final plans for USCENTCOM. Additionally, the structure of both the RDJTF and USCENTCOM played an important part in shaping the culture of the new unified command. This fact is readily apparent in the "culture of justification" that shaped commanders' interactions with their higher and peer organizations.[1] It also affected the focus of the new command.

BACKGROUND

The strategic importance of Southwest Asia has been recognized by the United States for decades. Unhindered access to the region's petroleum reserves has been a long-term U.S. strategic goal through a succession of Presidential administrations. These administrations relied on the "Twin Pillars" of the Shah of Iran and the Kingdom of Saudi Arabia to keep the region relatively stable.[2] The turbulent events of the 1970s fundamentally changed the foundation on which American policy was based, and propelled the region into the spotlight of continual presidential concern.

As the United States ascended to superpower status following World War II, in competition with the Soviet Union, the Middle East—along with policy for other regions of the world—came to be viewed through the lens of the Cold War. As such, the U.S. strategic focus in the region was on "denying turf and access to oil" to the Soviet Union.[3] In addition to DoD assets, the area was a major focus for the State Department, intelligence agencies, and many other executive agencies. Despite this history, no single command was assigned ultimate responsibility for synchronizing and directing operations or setting strategic objectives for the region. Indeed, the Middle East, broadly understood, was the responsibility of several different commands, including USEUCOM, USPACCOM, and USLANTCOM; all had responsibility for various portions of the region.[4]

This arrangement, while not perfect, was satisfactory until the 1970s when events in the region fundamentally changed the situation. Before, when the region was relatively stable, Southwest Asia was seen as an economy of force mission for the DoD, resulting in a collective American effort in the region that was uneven and uncoordinated. This arrangement was acceptable at the time because under the Richard Nixon Doctrine, the United States called for her allies to take an active part in countering threats in the region.[5] Throughout the 1970s, however, events in the "crescent of crisis" began to worry observers throughout the government. In the wake of the 1973 Arab-Israeli War, the JCS looked to new solutions for the region, proposing a "mobile task force" to respond to such crises, but the idea never gained any traction among decisionmakers.[6]

The overthrow of the Shah of Iran in 1979 fundamentally changed how the United States viewed the Middle East and led to recognition at the highest levels of the need for a unified way to deal with threats to American interests in the region. Additionally, the invasion of Afghanistan by the Soviet Union on December 26, 1979, made the prospect of Soviet troops in oil-rich Iran plausible. As a result of this rapid change in the strategic balance in the region, President Jimmy Carter announced what came to be known as the "Carter Doctrine" during his final State of the Union address in 1980. The Carter Doctrine stated that the oil fields of the Persian Gulf Region were "of vital interest to the United States, and that any outside attempt to gain control in the area would be 'repelled by use of any means necessary, including military force.'"[7] Subsequent to this announcement, the RDJTF was established at MacDill Air Force Base (AFB), Tampa, FL, on March 1, 1980.[8]

RAPID DEPLOYMENT JOINT TASK FORCE

The establishment of the RDJTF at MacDill AFB seems somewhat mystifying. Why would a Joint Task Force (JTF) with responsibility to react to crises in the Middle East have its headquarters in Florida? The answer lies not in the pleasant climate of Florida's West Coast, but in the original mission of the RDJTF that contributed to much of the organizational paranoia that has colored most of its existence and also that of its successor, USCENTCOM.

By design, a JTF is a temporary headquarters that is assembled for a finite period of time to deal with a specific situation. It is not a permanent organization and, by definition, is focused on the operational level of war. The result, in real terms, is that JTFs do not have organic troops or units. Instead, JTFs are essentially staff organizations that coordinate the efforts of component units from other commands and must communicate with the national command authority through their higher headquarters. The problem in the early days of the RDJTF was that it was designed to counter any contingency. As such, it was rather unfocused except for the mandate that it be able to deploy multiple division-sized elements to react to contingencies. The JCS came to the conclusion that the only command able to provide and transport the necessary forces was the U.S. Readiness Command (USREDCOM), that was based at MacDill AFB. Even with this particular issue resolved, support for the establishment of the RDJTF was far from unanimous. There was much debate within the JCS and Congress about which service should have primacy within the RDJTF. Several alternatives were proposed.[9] This interservice rivalry was a theme that continued even after the RDJTF was formally established under USREDCOM.

The first Commander of the Rapid Deployment Joint Task Force (COMRDJTF) was Marine Lieutenant General Paul X. Kelley. He was a well-respected officer who would later become the 28th Commandant of the Marine Corps.[10] He immediately conducted a series of command post exercises. These exercises demonstrated some practical problems with the current command arrangement. In addition to demonstrating some of the limits of the RDJTF structure, it also fueled Kelly's frustration with the CINC of USREDCOM, U.S. Army General Volney F. Warner.

The relationship between General Warner and his subordinate, Lieutenant General Kelley, was strained from the beginning and only worsened the longer they worked together. The relationship between the two men eventually deteriorated to the point of "personal animosity."[11] The relationship was strained for several reasons. First, Warner had advocated the return of responsibility for the Middle East to USREDCOM because in its former incarnation as U.S. Strike Command (USSTRIKECOM), it had been responsible for the region. Additionally, Warner also viewed correctly the RDJTF as a subordinate element of his command. Both of these views conflicted with the role that Kelley envisioned for the RDJTF. Kelley was particularly frustrated because he was not authorized to communicate directly with the JCS and the Secretary of Defense, independently of the CINC of USREDCOM, on matters relating to the RDJTF.[12]

The fact that Kelley was a Marine also adversely affected the relationship between the two commanders, especially because the competing versions of the proposed structure for the RDJTF gave primacy within the organization to different services. At the time, the proposal that seemed to have the most serious chance of implementation was the one that gave the Marine Corps primacy. Warner later recalled, "Unfortunately, we were both caught up in the service argument as to whether it should be a premier Army or Marine force."[13] The personal conflict between the two commanders eventually affected the members of their staffs. The environment at MacDill AFB was tense and confrontational, as the two units assumed the personality of their respective commanders. The situation got so bad that newspapers began to report on the animosity.

Then in April 1981, looking to overcome some of the issues that had plagued the RDJTF, the Secretary of Defense requested that the JCS formulate a plan to move responsibility for the Middle East from the JTF to a "separate unified command.[14] In response, the JCS established a three-phased plan for gradual transition over an 18-month period. In phase one, RDJTF remained subordinate to USREDCOM as component headquarters for the other services that were placed under operational control of COMRDJTF. During this time, Kelley was replaced by U.S. Army Lieutenant General Robert Kingston on July 17, 1981. Phase two would be concluded when RDJTF was removed from USREDCOM and activated as a separate command on October 1, 1981.[15] Phase three would conclude with the disestablishment of the RDJTF on December 31, 1982, and the establishment of USCENTCOM on January 1, 1983.[16] Kingston would then become commander of USCENTCOM.

U.S. CENTRAL COMMAND

A new name and upgraded status as a unified command did little to change the culture of justification that had permeated the RDJTF from its inception. The difference was that now the focus was to justify itself as a unified command on equal footing with the other unified commands. Despite claims to the contrary, the structure of USCENTCOM looked essentially the same on New Year's Day, 1983 as it had as RDJTF the previous evening. This fundamental stability made it easy for old habits, especially the focus of the operational level of war, to remain in the new strategic-level command, which had long-lasting effects on how USCENTCOM operated, and, some say, continues to operate even today.

When the RDJTF was disestablished and USCENTCOM was created with the same commander, it created a unique situation. Kingston had been the commander of a JTF, which was a three-star billet, but the job of commander of USCENTCOM was a four-star billet, putting Kingston in the unenviable position of being outranked by his peer CINCs. Undoubtedly, this situation did not go unnoticed by the other CINCs and the JCS. There is some indication that USCENTCOM was viewed by others as just the RDJTF with a new name.[17] USCENTCOM's official history states that "General Kingston's task as first Commander in Chief, United States Central Command . . . was to establish the command's *bona fides* as a unified command and as a credible force" and that "General Kingston championed the idea that USCENTCOM was a full-fledged unified command in exactly the same sense as the others."[18] Kingston was not promoted to General until November 6, 1984, a full 23 months after assuming command of USCENTCOM.[19] It is certainly reasonable to assume that his insecurity at his inferior rank was translated to his staff and manifested itself as organizational insecurity in regards to other unified commands.

In his effort to be seen as equal to the other unified commands and not just a renamed RDJTF, Kingston made significant structural changes, most of which were positive for the organization. He demanded that his command be assigned "actual component forces from the Services" and not "notional forces" as he had been assigned as COMRDJTF. He also eventually secured control of security assistance operations in the region from USEUCOM. This was extremely important because it was an important tool for strategic economic relations. Finally, he also assumed responsibility, from United States Air Forces in Europe of Airborne Warning and Control Systems.[20]

Despite all of the gains made by Kingston, when Marine General George B. Crist assumed command on November 27, 1985, he said:

> USCENTCOM was a unified command in name only . . . an RDJT whose sole purpose was to go to Iran and wage World War III against the Russians in a conflict restricted solely in our theater of operations.[21]

To counter this perception, Crist scrapped the operations plan for defending Iran against a Soviet invasion and put special emphasis on diplomacy in the region. He tried to convince others, and even went so far as to issue an order to his staff to start using the term "Arabian Gulf" to refer to the body of water commonly known as the Persian Gulf, out of consideration for the majority of the Arabian population in the region.[22] Despite his efforts, old habits die hard, and USCENTCOM was sucked down to the operational level of war in 1987 when USCENTCOM Naval forces began escorting reflagged Kuwaiti oil tankers in Operation EARNEST WILL, as well as a series of other incidents involving Iran.

The operational focus and culture of justification, left from the days of the RDJTF, can even be seen in U.S. Army General H. Norman Schwarzkopf's direction of Operation DESERT SHIELD and Operation DESERT STORM. Doctrinally, it would have been appropriate to assign a subordinate combined JTF to conduct these operations. This would have freed Schwarzkopf and his staff to remain focused on the strategic implications of the operations. Instead, he moved his headquarters forward to Saudi Arabia and directed the fight himself, then quickly returned his staff to MacDill AFB when the fighting was over. Because of this strict operational focus, he gave limited thought to the long-term effects of the operations on the region. Subsequent commanders of USCENTCOM have continued to follow this model. In fact, General Tommy Franks seems to have modeled his conduct of CENTCOM's response to the terrorist attacks on September 11, 2001, almost exactly as Schwarzkopf's. This model can be somewhat effective for short durations, but when the conflict becomes a long-term one, the model is less effective.[23] The commander and his staff become lost in the minutiae of operations (instead of leaving such matters to local commanders and thus remaining one level removed) and focused on setting conditions for long-term success in the region.

CONCLUSION

The fundamental change in the security situation in the Middle East in the 1970s caused the United States to have to come to grips with the fact that American national interests in the region were not being fully supported by the current military command. As a result, Carter created the RDJTF to try to create an organized way to respond to emergencies in the region. It became apparent quickly that inter- and intraservice rivalries, personal rivalries, and the temporary nature of a JTF made it difficult for the RDJTF to accomplish its mission. As a result, the creation of USCENTCOM was an organized and well-planned gradual process. It was designed to address the shortcomings of the JTF structure and provide a stable strategic headquarters to plan and direct U.S. efforts in the region. Unfortunately, the organizational culture of operational focus and the culture of justification that was central to the RDJTF remained at USCENTCOM and have affected how the command conducts operations to this day.

ENDNOTES - CHAPTER 4

1. Colonel David A. Dawson (USMCR), Command Historian, USCENTCOM, interview by the author on December 20, 2010, notes, Headquarters, USCENTCOM, MacDill AFB, FL.

2. Paul K. Davis, "Observations on the Rapid Deployment Joint Task Force: Origins, Direction, and Mission," Address to the 23rd Annual Convention of the International Studies Association, March 24-27, 1982, Santa Monica, CA: RAND, June 1982, p. 8, available from RAND Archive RAND/P-6751.

3. *Ibid.*

4. Anthony H. Cordesman, "USCENTCOM Mission and History," Washington DC: Center for Strategic and International Studies, 1998, p. 4.

5. Jay E. Hines, "From Desert One to Southern Watch: The Evolution of U.S. Central Command," *Joint Forces Quarterly*, Spring 2000, p. 42. For a good representation of many of the larger challenges shaping American strategy in the 1970s, see Figure 2, "Origins of current RDJTF readiness problems." in Davis, "Observations on the Rapid Deployment Joint Task Force," p. 7.

6. Jay E. Hines, *A Brief History of the United States Central Command*, MacDill AFB, FL: USCENTCOM History Office, February 1995, p. 3.

7. President Jimmy Carter in his State of the Union Address, January 23, 1980, quoted in Congressional Budget Office, "Rapid Deployment Forces: Policy and Budgetary Implications," Washington, DC: Congressional Budget Office, February 1983, p. 4.

8. Cordesman, p. 5.

9. Davis, p. 4. For an excellent discussion of the specific proposals and the underlying service rivalries involved, see Hendrick Bliddle, "Reforming National Security Organization: U.S. Military Command Arrangements for the Persian Gulf, 1977-1981," Master's thesis, University of Copenhagen, Copenhagen, Denmark, pp. 80-85, 209.

10. *"Who's Who in Marine Corps History,"* s.v. Paul X. Kelley, History Division, United States Marine Corps, available *https://www.mcu.usmc.mil/historydivision/Pages/Who%27s%20Who/J-L/Kelley_PX.aspx,* accessed April 2, 2015.

11. Dawson interview.

12. Congressional Budget Office, "Rapid Deployment Forces: Policy and Budgetary Implications," p. 4.

13. Interview with General Volney F. Warner, United States Army Retired, by Dean M. Owen, U.S. Army Military History Institute Senior Officer Oral History Program, U.S. Army War College, Carlisle, PA, 1983.

14. Ronald H. Cole *et al., History of the Unified Command Plan 1946-1996,* Collingdate, PA: Diane Publishing, 1996, p. 74.

15. Hines, *A Brief History of the United States Central Command,* p. 5.

16. Cole *et al.,* p. 74.

17. Dawson interview.

18. Hines, *A Brief History of the United States Central Command,* p. 6.

19. Colonel David A. Dawson (USMCR), Command Historian, USCENTCOM, "The Military in the Middle East World War II—Present," Power Point Presentation, November 12, 2009, p. 13.

20. Hines, *A Brief History of the United States Central Command,* pp. 6-7.

21. "End-of-Tour Report," General George B. Crist, *History of the United States Central Command*, Tampa, FL: USCENTCOM, 1988, pp. B1-B34, quoted in Hines, *A Brief History of the United States Central Command*, p. 8.

22. Dawson, "The Military in the Middle East World War II—Present," p. 15.

23. Colonel David A. Dawson (USMCR), "The Evolution of U.S. Central Command From Operational To Strategic Headquarters," Masters of Strategic Studies, Thesis, United States Army War College, Carlisle, PA, 2010, p. 1.

CHAPTER 5

OVERCOMING INERTIA THROUGH SIMULATION: U.S. TRANSPORTATION COMMAND

Gail E. S. Yoshitani

This essay was directly adapted from a master's thesis written by the author for Duke University, Durham, NC, and completed in 2001. While large portions of this work are directly taken from that thesis, for ease of reading, they are not so marked. The full thesis is available from Defense Technical Information Center, *www.dtic.mil/get-tr-doc/pdf?AD=ADA412949.*

INTRODUCTION

The military leaders serving on the Joint Chiefs of Staff (JCS) in the late-1970s understood that the nation's ability to quickly and decisively relocate forces and material was a linchpin in the nation's defensive strategy of deterrence. To that end, they understood the importance of responsiveness in the nation's military transportation system to the direction of the National Command Authority (NCA). They demonstrated that understanding by undertaking a study to identify and evaluate which system of military command and control would ensure the best responsiveness to the direction of the NCA in times of crisis and war. Nevertheless, in 1977, the Joint Chiefs chose to retain a flawed transportation system because considerations, organizational power structures and relationships, and other bureaucratic concerns—rather than warfighting—focused their decisionmaking. The primary reason those bureaucratic issues were able to drive decisionmaking was that they were more familiar and tangible to military leaders than warfighting concerns that remained quite intangible and difficult to conceptualize. In 1978, a computer simulation, code-named NIFTY NUGGET, led the Joint Chiefs to change course and create a new coordinating authority for mobilization deployment planning, the Joint Deployment Agency (JDA), which was a forerunner to the United States Transportation Command (USTRANSCOM). This chapter demonstrates that the power of simulation exercises stem from their ability to provide an impetus for positive change by enabling military leaders to realistically confront their systems of operation and make decisions for improvement based on what is best for warfighting.

The years between 1978 and 1987 witnessed a profound transformation of the U.S. ability to project force. At the beginning of this period, the nation's method for moving soldiers and equipment consisted of a patchwork of Army, Navy, and Air Force transportation systems, organized and effective on paper but disorganized and ineffective in reality. Although they all fell under the purview of the Department of Defense (DoD), each service's transportation system nonetheless remained divided into separate domains that were stubbornly isolationist in their dealings with one another, and only fitfully and infrequently drawn together as a unified force. However, with the formation of the U.S. Transportation Command (USTRANSCOM) in 1987, these rivalries were superseded by unified imperatives. Charged with providing global air, sea, and land transportation to meet national security objectives, the new command placed a single leader in charge of all three services' transportation systems, thus beginning a dramatic conversion from disjointedness into jointness.

On April 18, 1987, President Ronald Reagan signed the order that directed Secretary of Defense Caspar Weinberger to establish USTRANSCOM. The order represented victory for some and defeat for others, for all attempts to unify DoD transportation functions had met with resistance. This opposition was ultimately overcome in 1986 by a three pronged assault: the first came from Reagan's Blue Ribbon Commission on Defense Management, known as the Packard Commission, that recommended establishing a unified command; the second came from the Goldwater-Nichols Defense Reorganization Act of 1986, that ordered the Secretary of Defense to consider the creation of a unified transportation command; and the third came from Air Force Lieutenant General Alfred G. Hansen, Joint Staff Director of Logistics, who convinced the Chairman of the Joint Chiefs of Staff (CJCS), Admiral William J. Crowe, to recommend to the President the establishment of the unified transportation command over the nonconcurrences of the Chief of Naval Operations (CNO) and Marine Corps Commandant. These matters are related in a number of oral histories conducted by Dr. James K. Matthews and Dr. Jay H. Smith, in their roles as Directors of the USTRANSCOM Research Center.[1]

While each of the aforementioned efforts served as important stimuli in prompting the development of USTRANSCOM, this chapter focuses on an earlier period in time and on a particular stimulus that proved capable of overcoming enough bureaucratic inertia to enable the JDA, a forerunner to USTRANSCOM, to be formed. That stimulus was known as NIFTY NUGGET, a simulation exercise run in October 1978 by the JCS. Truly, change does not come easily to the DoD; however, this story demonstrates that simulation can help leaders in a large organization, such as the military, change their corporate minds and strike out in new and uncomfortable directions. The power of simulation exercises such as NIFTY NUGGET stems from their ability to enable military leaders to confront realistically their systems of operation and make decisions about what is best for warfighting, thus providing an impetus for positive change.

NIFTY NUGGET 78

According to a JCS memorandum, NIFTY NUGGET 78 was designed to "test Service and joint plans and procedures during full mobilization and initial deployment processes."[2] The operation itself was 3 weeks long, running from October 10 to October 31, 1978. Participants, representing 52 different DoD and Federal Civil Departments and Agencies, reacted to computer-generated scenarios just as they would during actual mobilization and deployment for war.[3] NIFTY NUGGET was the first time in the nation's history that an exercise of such magnitude and specific focus was attempted. The baseline scenario for the exercise was a fast-breaking attack by Warsaw Pact forces on North Atlantic Treaty Organization (NATO) forces in Europe, a contingency that leaders in the DoD considered one of the least likely, but most dangerous, of Cold War scenarios.[4] The exercise synopsis accurately reflected the heart of all American defense planning during the Cold War: the defense of Western Europe.

Exercise participants quickly discovered significant problems in mobilizing, deploying, and sustaining American military forces. While Soviet forces marched through one European country after another, U.S. military leaders encountered severe difficulties transporting combat soldiers, supplies, and equipment to Europe. Some units were flown into the combat zone without all their personnel because they had been left behind by the Air Force. Such units were forced to fight shorthanded, some for as long as 4 days, before the Air Force finished transporting the 14,304 bypassed soldiers.[5] One account of the simulation exercise suggests that close to 400,000 American Soldiers in the theater of operations "died" because they failed to receive needed supplies.[6] None of the services was blameless. The military transportation system of planes, ships, trucks, and trains to support military defensive plans in Europe had proven unable to carry out its mission.

Fortunately for the nation, the Soviet attack and failed response took place only in the processors of a DoD computer. Instead of a disaster for the United States and its NATO allies, the simulation was a profound and ominous warning that the nation was wholly unprepared to defend Europe, largely because there was no single manager to oversee the mobilization and deployment of forces.

While the JCS planned NIFTY NUGGET to be an exercise primarily focused on mobilization, the problems of deployment and defense department transportation operations soon overshadowed the design. An analysis report from the joint staff on the exercise confirms that "a shift in exercise focus from mobilization to deployment seemed to occur."[7] The sheer volume of reports detailing problems in the deployment phase of the exercise indicates that a large number of participants viewed this as very important and helped to pinpoint numerous defense transportation shortcomings. Interest in this aspect of the exercise is unsurprising given that the military's ability to quickly and decisively relocate forces and material was a linchpin in the nation's defensive strategy of deterrence. Frighteningly, NIFTY NUGGET made that linchpin appear very suspect. By definition, a linchpin is a "central and cohesive element," but in 1978, while many government leaders might assert that the nation's defense transportation system was **central** to securing national security interests, one would have had difficulty finding leaders who would declare the system **cohesive**.[8]

The Problem: "The Single Manager Plan."

During the NIFTY NUGGET exercise, participants relied upon a military transportation system that had been in place with very little change since it was first developed in 1955 by the Office of the Assistant Secretary of Defense. Although it was a system designed to minimize inefficiencies across the three services' (Army, Navy, and Air Force) transportation systems, it also was meant to maintain peace between the services. Ironically, the plan that won approval in 1955 was called the "Single Manager Plan," so named because it placed each service in charge as the "single manager" of transportation assets in its area of expertise, but this effectively produced multiple "single managers." Although the Secretary of Defense was responsible for the overall management of defense transportation issues, he delegated that responsibility to his three service secretaries. In turn, each service Secretary further delegated responsibility to an executive agent of a Transportation Operating Agency (TOA).[9] For the Army, the executive agent was the Commander of Military Traffic Management Command (MTMC); for the Navy, the Commander of Military Sealift Command (MSC); and for the Air Force, the Commander of Military Airlift Command (MAC). Respectively, each TOA was assigned responsibility of its service's general area of expertise: MTMC was designated the single manager for military traffic, land transportation, and common-user ocean terminals; MSC was the single manager in charge of common-user sealift; and MAC was the single manager in charge of airlift services.

Although the breakdown of responsibility seemed straightforward, the command relationships established under the Single Manager Plan for each TOA were far from simple. Each TOA was technically commanded by its parent service, but its single manager status made it ultimately responsible to the DoD as a whole. Consequently, each TOA received guidance from both its service secretary and from the Office of the Assistant Secretary of Defense (Manpower, Reserve Affairs and Logistics) (OASD MRA&L). MAC's designation as a specified command on February 1, 1977, added yet another direct link within its command structure to the JCS. Additionally, MTMC's designation as traffic manager for the DoD produced a very peculiar relationship for that organization with other services. In its capacity as traffic manager, MTMC was charged with overseeing and evaluating the entire DoD transportation system that awkwardly included inspecting areas within the other services' jurisdiction. Finally, each TOA relied on numerous connections with civilian organizations outside of the DoD to ultimately determine its success or failure.[10]

In 1979, the OASD MRA&L contracted the Harbridge House, a think tank headquartered in Boston, MA, to analyze the "functional and organizational interrelationships of the three transportation operating agencies."[11] When the Harbridge House published its findings, it described the defense transportation system as "characterized by splintered responsibilities and initiatives, fractionated and incompatible systems, and divided loyalties and interests."[12] The report further emphasized that:

> such disparity of interests is evident in entrenched parochialism, inadequate and incorrect documentation, inefficient and duplicative procedures, added costs, and a limited ability to respond to national command authorities.[13]

Remarkably, this same system managed, in spite of such shortcomings, to function under peacetime conditions. Nevertheless, the problems caused by the complexity of relationships born from the Single Manager Plan developed in 1955 were clearly revealed during the military scenarios executed during NIFTY NUGGET.

The deployment management problems identified during NIFTY NUGGET seemed to stem from a single root cause—the absence of a single manager for mobilization and deployment. This conclusion was not lost on the Senior Observer Group tasked with evaluating the exercise; it was immediately clear to them that issues of command and control were negatively affecting the overall efficacy of the defense transportation system. Thus very early in the exercise, they recommended to CJCS General David C. Jones that he "designate one headquarters or agency responsible for all deployments."[14] The observers advised the Chairman that such an agency should be responsible for managing all overseas movements, maintaining the databases that support such movements, and coordinating deployment activities between the Joint Staff, commands, and separate TOAs.[15]

Remarkably, a first version of the new terms for an agency to serve as the single manager for deployment planning was forwarded to the JCS by November 17, 1978, just 17 days after the conclusion of the NIFTY NUGGET exercise. General John J. Hennessey, Commander of the United States Readiness Command (USREDCOM), was assigned the task of overseeing the creation of the new agency. Hennessey's final plan for the new agency incorporated general recommendations from each service's transportation management agency and comments from the supported unified and specified commanders. By March 27, 1979, the JCS had approved the terms creating the JDA, and on May 1, 1979, the organization was officially established.[16] Its mission was to serve as the Joint Chiefs' coordinating authority for mobilization deployment planning.[17] The JDA was located at MacDill Air Force Base in Tampa, FL, along with the USREDCOM.[18]

Developing the JDA moved the Joint Staff in a new and uncomfortable direction; it was unprecedented for combatant forces to be placed under the direction of an agency and not a command.[19] The military chain of command ran from the President to the Secretary of Defense to the unified and specified commanders. Neither the military departments nor the JCS were included in that chain of command. The military departments were responsible for training and supplying the soldiers, sailors, airmen, and marines who were assigned to the unified and specified commanders, while the JCS served as military staff and advisors to the Secretary of Defense. Therefore, it is not surprising that the initial terms developed for the JDA made it only a coordinating authority. Nevertheless, the JDA's initial designation as a coordinating authority should not be seen as a slight.[20] Military leaders had taken a first step in an uncharted direction by inserting this new agency into their organizational system.

The Power of Simulation.

While the establishment of the JDA was an important milestone along the path toward the creation of the U.S. Transportation Command (USTRANSCOM), the primary reason for looking at the JDA's formation is to see how the Joint Chiefs came to conceptualize their transportation problems and how they established their willingness to solve those problems by moving in new directions. Thanks to NIFTY NUGGET, Jones and the other members of the JCS identified 487 deficiencies that required correction in the nation's system of mobilization and deployment for warfare or emergency operations. While not all of those deficiencies were related to the transportation system, those that did were identified as significant problems.

The logic and simplicity of the reforms the JCS implemented in 1978 begs the question: why was such an agency not created earlier? As it turns out, the Joint Chiefs had considered just such an agency in June 1977, just 15 months before the start of NIFTY NUGGET. At that time, however, the Chiefs elected to maintain the status quo, concluding that, "no major deficiencies were identified within the current peacetime and wartime" transportation structure of command and control and "no further organizational realignments . . . should be undertaken."[21]

Why did the military leaders serving on the JCS, responsible for ensuring that the nation's armed forces were prepared to deploy around the world, decide in 1977 to keep a military transportation system that proved to be so flawed in 1978? The answer lies in how the JCS assessed both the existing transportation system, which they chose to retain, and the study alternative that, if selected, would have created an agency akin to the JDA. The Joint Chiefs' justification for why the current transportation system was preferable to the other study alternatives provides a clear contrast between how they conceptualized these issues with and without the benefit of simulation. NIFTY NUGGET was decisive in changing how the JCS conceptualized the issues surrounding defense transportation.

The purpose of the Joint Chiefs' 1977 study was to analyze 10 alternative methods for command and control over the services' transportation managers. A cover memorandum from the Joint Chiefs to the Secretary of Defense submitted with the study's results explained that the goal of the study was:

> to identify and evaluate alternatives which would insure [sic] responsiveness to direction to by the NCA [National Command Authority] in times of crisis/war and compatible peacetime economies in procurement, management, and resource utilization, with consideration given to maintaining an appropriate balance between unified and Service needs.[22]

It further specified that the Joint Staff had analyzed 10 alternatives, each in terms of its "responsiveness to unified direction, command relationships, economy and efficiencies, operating procedures, Service requirements, funding, and legality."[23] Nevertheless, the Joint Chiefs specifically pointed out in their memorandum that "crisis and wartime responsiveness to the NCA was the primary criterion" used in their evaluation process and that "compatible peacetime economies were a secondary criterion."[24]

The Joint Chiefs selected the study alternative advocating continuance of the current system. Under that system, the Air Force's transportation management agency, MAC, was a specified command, while the other two services' transportation agencies, the Navy's MSC and the Army's MTMC, received direction from their respective services. MAC was designated a specified command on February 1, 1977, taking the agency out from under the command and control of the Secretary of the Air Force and placing it directly under control of the JCS during war. This command relationship enabled it to operate and plan airlift matters directly with the other unified commanders.[25] While that arrangement protected MAC from receiving guidance from multiple channels, the other two services' transportation managers were still subject to receiving direction from both their individual services and from the JCS. It was this system, in which each service's transportation manager received guidance from different sources, that ultimately produced many of the problems encountered during NIFTY NUGGET.

According to the Joint Staff report, MAC was accorded its special command relationship because it controlled significant "forces in peacetime as well as wartime," while the Army and Navy's transportation managers were viewed basically as "managers of contractor assets."[26] The report argued that airlift was more likely to be required on short notice for contingencies and crises than surface lift, the movements by trucks and rail contracted by the Army and by ships contracted by the Navy.[27] Nonetheless, the evaluators did note that sealift carried 90 percent of the supplies delivered to U.S. forces in both the Korean and Vietnam wars. Furthermore, although Army and Navy transportation managers relied on contracting civilian assets for movements, they still played a large role in the success of transportation of soldiers and equipment. After all, it was the Army that was ultimately responsible for the arrival of all soldiers and equipment at the air or sea point of embarkation and the overseas ports of debarkation.[28]

Interestingly, the Joint Chiefs did note in their findings that the one disadvantage of selecting to maintain the status quo was that it did not enhance responsiveness of the MTMC or the MSC to unified direction or coordination.[29] Although the purpose of the study was to identify a system of command and control that would ensure responsiveness to the direction of the NCA over the services' transportation managers, this disadvantage did not weigh heavily against maintaining the status quo. The Joint Chiefs reconciled this conflict by explaining, in a memorandum to the Secretary of Defense, that while responsiveness to the NCA might be improved by establishing a direct organizational relationship to the JCS, no major deficiencies had been uncovered in the current organizational structure to warrant such realignment. They reasoned further that "unless an in-depth cost-benefit analysis indicated significant projected long-range saving" no further organizational realignments needed to be undertaken.[30]

The following points can be made with the knowledge that a mere 15 months later during NIFTY NUGGET, significant problems were uncovered. There is little question that the Joint Chiefs believed that the military transportation system needed to be responsive to the NCA, a point emphasized numerous times throughout their 1977 study results. However, there is a question as to whether they had an accurate conception of this relationship before their NIFTY NUGGET experience. Examining the Joint Chiefs' criticisms of the study alternative that would have created an agency similar to the JDA provides a look at how they conceptualized transportation issues without the benefit of simulation.

Under this particular study alternative, all three services' transportation managers, even the MAC, would be responsive to a transportation agency that would report to the Secretary of Defense through the JCS.[31] The Joint Chiefs cited five specific disadvantages to this alternative.

1. The Joint Chiefs did not like the idea of adding an additional organizational element into the chain of command between the Joint Chiefs and the transportation managers.

2. The Joint Chiefs were against the idea of removing the military department secretaries from their historical role as single managers over their service's transportation assets.

3. While the Joint Chiefs agreed that having one single manager for transportation did seem advantageous under this organizational structure, the additional headquarters would require increased manning without decreasing any manning requirements at the service levels.

4. This plan would place the JCS between the Secretary of Defense and the service transportation agencies, a role to which they were unaccustomed.

5. The Joint Chiefs were wary of making the unprecedented move to place combatant forces under the direction of an agency and not a service.[32]

Clearly, NIFTY NUGGET revealed those five concerns to be less consequential than the need for coordination. Yet, in 1977, the JCS did not have the benefit of that experience to guide their thinking and conceptualization of these transportation issues. Some will argue, nevertheless, that as senior military leaders responsible for ensuring that the nation's armed forces are prepared to deploy around the world, they ought to have been able to envision such problems without a simulation exercise. Such arguments, however, do not change the fact that for some reason, in 1977 the leaders serving on the Joint Staff, although stating that warfighting was their primary criterion, placed a much greater focus on bureaucratic concerns. Economic considerations, political agendas, and organizational power structures and relationships ruled the day. Then, for at least a short time in 1978, this bureaucratic noise was muted by the NIFTY NUGGET simulation exercise that demonstrated to military leaders in a stark and frighteningly realistic way the strengths and weaknesses of their current system of operations. When forced to confront these problems head on, warfighting concerns moved to the forefront and bureaucratic concerns became simply hurdles to overcome.

It may be also argued that the Joint Chiefs based their decision for the status quo upon a desire to satisfy internal political objectives. In other words, they recognized that maintaining the status quo was not the best alternative but accepted it to avoid organizational strife. Interservice rivalry should always be taken into consideration when examining and assessing military decisions and activities, for no service wants to relinquish dollars, responsibility, or authority to another. Historically, many fierce battles have been waged between the three services over such issues. As discussed in the beginning of this chapter, the transportation system as it stood in 1977 was one that had been pieced together by the Office of the Assistant Secretary of Defense in 1955 to maintain peace between the services by splitting responsibilities for transportation between all of them.

This chapter asks readers to accept several judgments: first, that the leaders serving on the JCS in 1977 recognized the importance of the nation's military transportation system's responsiveness to the direction of the NCA; second, that even though those leaders recognized that their current system of command and control did not optimize service responsiveness, they still considered it to be the best in meeting all considerations; and last, that the primary reason military leaders retained the status quo stemmed from their inability to accurately conceptualize warfighting issues that left bureaucratic issues to drive their decisionmaking.

CONCLUSION

The military leaders serving on the JCS in 1977 understood the importance for the nation's military transportation system to be responsive to the direction of the NCA. They demonstrated this understanding not only by undertaking a study to identify and evaluate which system of military command and control would ensure the best responsiveness to the direction of the NCA in times of crisis and war, but also by their statements made throughout that study regarding the importance of responsiveness.

In their study, they determined that their current system of command and control, while not maximizing all services' responsiveness to the NCA, was still the best option of those considered. They disregarded one alternative that would have created an agency similar to the JDA, citing concerns over adding an additional organizational element into the chain of command, removing military department secretaries from their role as managers over their services' transportation assets, adding a requirement for more manpower, positioning the Joint Staff in between the Secretary of Defense and service transportation agencies, and placing combatant forces under the direction of an agency and not a service.

NIFTY NUGGET identified serious deficiencies in the military transportation system that the Joint Chiefs were relying upon to carry out the security strategy of the nation, one that relied heavily upon conventional forces being able to rapidly deploy around the world. Once those shortcomings were identified, the Joint Chiefs immediately set out to implement changes in their operating system.

This chapter has argued that the Joint Chiefs retained a flawed transportation system in 1977 because economic considerations, organizational power structures and relationships, and other bureaucratic concerns focused their decisionmaking rather than warfighting concerns. The primary reason those bureaucratic issues were able to drive decisionmaking was that they were more familiar and tangible to military leaders than warfighting concerns, which remained quite intangible and difficult to conceptualize. NIFTY NUGGET led the Joint Chiefs to change course and create a new coordinating authority for mobilization deployment planning, the JDA, which was a forerunner to USTRANSCOM. This chapter shows that the power of simulation exercises stem from their ability to provide an impetus for positive change by enabling military leaders to realistically confront their systems of operation and make decisions for improvement based on what is best for warfighting.

ENDNOTES - CHAPTER 5

1. For instance, see Dr. James K. Matthews, *General Alfred G. Hansen, United States Air Force (Retired): Establishment of the United States Transportation Command,* Scott Air Force Base (AFB), IL: USTRANSCOM, March 1999; Dr. Jay H. Smith and Margaret J. Nigra, eds., *Reflecting on the Origins of the United States Transportation Command: A Panel Discussion,* Scott Air Force Base, IL: USTRANSCOM, October 4, 2007; Dr. Jay H. Smith, *Lieutenant General Edward Honor, Establishment and Evolution of the United States Transportation Command,* Scott AFB, IL: USTRANSCOM, August 2006.

2. The *Joint Chiefs of Staff memorandum* (JSCM)-*465-77,* dated January 10, 1978, sent to the Secretary of Defense, Subject: Exercise NIFTY NUGGET 78 (U), Scott AFB, IL: USTRANSCOM Research Center, p. 1. The research center holds all USTRANSCOM archival material as well as the retired historical reports and archives from the JDA, 1979-87. The archive will be abbreviated as TCRC. Note that not all sources will have a file and record number because some have come from the archive's library and safes. This particular memorandum also provided a brief history of the origins of the exercise. The memorandum states that a memorandum from the Secretary of Defense, December 23, 1976, "Mobilization and Deployment Planning Guidance," directed the JCS and the Services to make plans for an "extensive testing of the full mobilization process." By August 1977, the Office of the JCS informed the chiefs of the services and the commanders of the unified and specified commands that a joint mobilization exercise would be run in October 1978. The concept and objectives for Exercise NIFTY NUGGET 78 were published as an attachment to JCSM-465-77.

3. *Detailed Analysis Report Exercise Nifty Nugget 78 (U),* April 11, 1979, prepared by Operations and Exercise Analysis Branch, Exercise Plans and Analysis Division, Operations Directorate (J-3), OJCS. "Executive Summary," EX-1, TCRC. The TCRC archive also has an extensive collection of documents on NIFTY NUGGET. Henceforth this particular source document will be abbreviated in the following manner *Nifty Nugget 78 (U).*

4. Harold Brown, Department of Defense Annual Report, Fiscal Year 1979, Washington, DC: Government Printing Office, 1978, p. 78.

5. *Nifty Nugget 78 (U),* "Executive Summary," EX-11; *Nifty Nugget 78 (U),* Section III-Deployment Processes, pp. III-24, III-26, III-27.

6. James K. Matthews and Cora J. Holt, *So Many, So Much, So Far, So Fast: United States Transportation Command and Strategic Deployment for Operation Desert Shield/Desert Storm,* Scott AFB, IL: Joint History Office, Office of the Chairman of the Joint Chiefs of Staff and Research Center, USTRANSCOM, 1996, p. 1.

7. *Nifty Nugget 78 (U),* Section III-Deployment Processes, p. III-16.

8. *The American Heritage Dictionary: Second College Edition,* s.v. "Linchpin."

9. *Historical Background: Military Traffic Management Command,* Washington DC: Military Traffic Management Command, 1985, p. 13, TCRC.

10. Harbridge House, *A Study of DoD Organization for Transportation and Traffic Management,* Boston, MA: Harbridge House, 1980, p. III-2, file 32.A-6-A, RG 32, TCRC. Narrative is drawn heavily from this source.

11. *Ibid.*, p. I-1.

12. *Ibid.*, p. III-17.

13. *Ibid.*

14. *Nifty Nugget 78 (U),* Section III-Deployment Processes, p. III-9.

15. *Ibid.*

16. Joint Deployment Agency History (U) 1979, p. 1, TCRC.

17. The organization's official history explains that "the term 'mobilization deployment planning' means the act of using authorized systems and measures for planning, coordinating, and monitoring movements and deployments of mobilized forces and material necessary to meet military objectives." See *ibid.*

18. In September 1977, President Carter requested that the Secretary of Defense conduct a review of the National Military Command Structure. That study recommended that the USREDCOM "be designated as the focal point for the coordination of the day-to-day aspects of mobilization/deployment planning of all CINCs, particularly as they pertain to lift requirements and detailed follow-through during major reinforcements." See *The National Military Command Structure: Report of a Study Requested by the President and conducted in the Department of Defense*, Washington DC: Government Printing Office, 1978, pp. 20-21.

19. JCSM-264-77, "Appendix: Report on the Analysis of Alternatives for Control of the Military Airlift Command, The Military Sealift Command, and the Military Traffic Management Command,", file 32-A-3-A, RG 32, p. 33, TCRC.

20. Joint Deployment Agency History (U) 1981, Vol. II, TCRC. In October 1981, the agency was given new terms increasing its power and authority.

21. JCSM-264-77, dated June 15, 1977, "Appendix," file 32-A-3-A, RG 32, p. 12, TCRC.

22. JCSM-264-77, file 32-A-3-A, RG 32, p. 1, TCRC.

23. *Ibid.*

24. JCSM-264-77, "Appendix," file 32-A-3-A, RG 32, p. 2, TCRC.

25. *Ibid.*, pp. 13-16.

26. *Ibid.*, p. 11.

27. *Ibid.*

28. *Ibid.*

29. *Ibid.*, p. 15.

30. JCSM-264-77, file 32-A-3-A, RG 32, p. 2, TCRC.

31. JCSM-264-77, "Appendix," file 32-A-3-A, RG 32, p. 5, TCRC.

32. *Ibid.*, pp. 32-35.

CHAPTER 6

COMMANDING THE FINAL FRONTIER: THE ESTABLISHMENT OF A UNIFIED SPACE COMMAND

Samuel P. N. Cook

INTRODUCTION

In 1974, the U.S. Joint Chiefs of Staff (JCS) circulated a memorandum to service chiefs asking for their opinion regarding the creation of a unified space command (U.S. Space Command [SPACECOM]). The chiefs uniformly rejected the proposal and argued against "militarizing" space. Yet the chief opponent of a unified space command at the time, the Air Force, soon became its biggest advocate and took the lead in establishing an Air Force Space Command (AFSPC) that would serve as the centerpiece for SPACECOM that was formed when President Ronald Reagan announced his Strategic Defense Initiative (SDI). The story of the establishment of the SPACECOM presents a cautionary tale about the pitfalls of establishing a military architecture in a theater or technological space in which none has existed before. This chapter describes the formation of SPACECOM, focusing on its decisions on the best way to support national defense and avoiding the adverse effects of appealing to narrow service interests during its critical formative period.

In 1974, commander of the U.S. Air Force Aerospace Defense Command (ADCOM) General Lucius D. Clay, Jr., wrote a letter to the Chairman of the JCS (CJCS) advocating the creation of a "Space Command." The CJCS, General David C. Jones, took this recommendation and wrote a letter to the head of each of the military branches and asked them to provide their opinions on the creation of such a command. The responses from the Chiefs presented a uniformly skeptical response to the idea of turning space into a military field of operations. The Air Force leadership especially opposed the formation of such a command because it was already the funding component for space and did not want to divert resources away from its primary mission of achieving strategic and tactical air superiority. The Chiefs were hesitant to divert time and resources to the potentially expansive project of developing space into a military theater of operations, especially given their current focus on disengaging from Vietnam and an increasing fear of the potential for Soviet space-based anti-missile capabilities. Just over 10 years later, on September 23, 1985, CJCS General John W. Vessey, Jr., presided over the formation of an organization that would, in fact, establish space as a theater of war: SPACECOM at Peterson Air Force Base (AFB) in Colorado. In his remarks, he stated that the new Space Command, "is not a force built to escalate the arms race. . . . The command will make its contribution to that fundamental element of United States strategy, the prevention of war."[1]

Less than 10 years after it had uniformly rejected the idea of making space into a theater of operations, the Department of Defense (DoD) fully embraced the concept of space as a new theater of military operations. The formation of SPACECOM presents a classic case study of how the DoD dealt with forming a new theater of operations based on exploiting and developing new technologies. But while space became a new theater of operations, DoD never envisioned it as a literal battlefield or a tactical objective. Instead, they viewed it as a new dimension—a technological space that would enable and enhance operations in the geographically based unified commands that were in place and served as the central fronts in the Cold War.

The formation of SPACECOM went through three phases of development. Between 1974-81, a small cadre of U.S. Air Force staff officers conceptualized and developed a vision for an AFSPC, but they ran into the strong headwinds of downsizing, given the Jimmy Carter administration's reluctance to escalate the Cold War. In 1979, the Air Force leadership supported the creation of an AFSPC, which became operational in 1982 after the inauguration of Ronald Reagan as President. This administration initiated a key component of its vision of the Cold War by launching the SDI, popularly known as "Star Wars," and under this umbrella the AFSPC quickly became a reality. As it became clear that space-based technology would become the foundation of a new integrated strategic posture, the AFSPC pushed for the creation of a fully integrated SPACECOM. A unified command would combine the capabilities of all services, would allocate Title X funding to space operations, and would grant appropriate command authority to influence national defense policy. The process of actually creating a unified command lasted from 1982-84, as DoD grappled with how to organize and integrate all space-based capabilities across the services. The concept of a SPACECOM ran into many challenges on the road to its formation: the challenge of merging assets from disparate commands from all sources, properly defining its new mission in an unknown technological and spatial environment, and finally dealing with interservice rivalry.

DEVELOPING A VISION FOR SPACECOM

From 1974-1981, advocates of turning space into an arena for military operations faced a significant series of challenges. First, the operational experience of most of the Pentagon leadership had been formed during the Vietnam War, and they saw little that space technology could have done to help in Vietnam. Moreover, the budget cuts and troop reduction of the Vietnam era became a priority among the Pentagon leadership. Finally, the Carter administration pushed to streamline and reduce command structures during this period, making the case for creating a new command significantly more difficult. But despite these prevailing headwinds, the advocates for space operations forged ahead with their vision and even used these challenges to shape the future of a SPACECOM.

On July 1, 1975, the JCS designated ADCOM as a specified command reporting directly to the JCS for operational control; on the same day Continental Air Defense Command (CONAD) was deactivated.[2] ADCOM now assumed all responsibilities of CONAD. Sensing that its mission was also becoming obsolete, planners within ADCOM sought to redefine its role around space operations. Despite the original opposition in the JCS towards expanding into space, a group of talented staff officers within ADCOM developed a vision for a future SPACECOM that would provide the foundation of future space operations under SDI. In October 1976, ADCOM produced a memorandum outlining the case for a space defense command. *Required Operational Capability 5-76 Memorandum* called for establishing a defense operations center that would serve as an all-source command and control center.

This advanced space command center would control anti-satellite operations, execute a plan for survival in the event of a nuclear war, control space surveillance, and conduct operational functions related to wartime operation of satellite ground stations. This new planning memorandum marked a shift to an aggressive space posture.[3]

In 1977, the Carter administration took control at the Pentagon and made a large push to downsize and streamline the military to deal with rising combat budget deficits and the nation's overall economic troubles in the aftermath of the Vietnam War. Air Force Chief of Staff General David C. Jones authored a study that proposed combining North American Air Defense Command (NORAD) and ADCOM. Both organizations advocated the move and planned to transfer ADCOM aircraft to Tactical Air Command, missile warning systems to NORAD, and space capabilities to Strategic Air Command (SAC). Secretary of the Air Force Harold Brown approved disestablishing ADCOM on November 20, 1977. A group of officers at ADCOM, Operations and Plans division, believed that the decision to deactivate ADCOM would sacrifice the hard-earned knowledge and expertise they had developed in advancing space operations. They feared that the body of knowledge gathered in Colorado Springs, CO, would be dispersed with the disestablishment of the command. Lieutenant Colonel Earl Van Inwegen led a group of junior officers who produced a series of studies and presentations over the next 5 years to keep the idea of space operations alive.[4]

In 1979, Air Force leadership made a dramatic shift and began to embrace the idea of space operations that the planners at ADCOM had long advocated. In 1978, for example, Secretary Thomas C. Reed commissioned a Space Mission Organization and Planning Study (SMOPS). On February 5, 1979, SMOPS published its recommendations. It formally advocated creating a space command within the Air Force and revising *DoD Directive 5160.32* that recommended that the Air Force receive the designation of DoD Executive Agent for Space. The Air Force quickly realized that it needed to develop a space operations organizational structure.[5] On March 1, 1979, the Assistant Secretary of Defense directed the Air Force to prepare a plan establishing Space Defense Operations Center (SPADOC) responding to the *Required Operational Capability 5-76 Memorandum*. He asked for a plan by April 1 and an established SPADOC by July 1 of that year, reflecting the urgency of the mission. SPADOC met these deadlines and officially began operations on October 1. At the same time, the Air Force enhanced NORAD's role in space and announced plans to deactivate ADCOM. Ultimately, it turned out that the requirement of downsizing benefitted the Air Force as it forced the service to consolidate and improve efficiency, both essential foundations for the formation of an operational space command.

U.S. AIR FORCE SPACE COMMAND

By 1979, Air Force leadership was thoroughly behind the concept of establishing a space command capability, but the Carter administration did not encourage this vision. In an effort to downgrade the missile defense mission, the administration downgraded the commander of NORAD and ADCOM to a three-star billet. The staff at ADCOM also faced an overall hostile environment due to severe budget cuts. Soon the Air Force decided to consolidate the Satellite Test Center (STC) at Fort Peterson in November 1979.[6] Again, these significant external pressures actually served a positive end, as they forced the Air Force to clarify its vision and to streamline space operations. Once Reagan took office in 1981, these painful but necessary moves catapulted the Air Force space advocates to the forefront of national security policy development and implementation.

In March 1979, Secretary of the Air Force John C. Stetson hosted a conference at the U.S. Air Force Academy in Colorado Springs, and ADCOM operations and plans officers had developed a briefing for exactly this occasion, one designed to push for saving ADCOM and orienting it towards the emerging mission of space defense and to expand it into a full space command. Privately, the planners called it the "Save the Day" briefing. At the last minute, Major General Bruce Brown pulled Van Ingwegen aside and told him they were "beating a dead horse." The staff shelved the presentation and renamed it the "dead horse briefing."[7] It was never given. However, in 1980, Lieutenant General Jerome F. O'Malley, the deputy Chief of Staff for operations, plans, and readiness, directed the Air Force staff to begin planning for space defense. Van Inwegen went back to the shelf and pulled out the "dead horse briefing." Then, after Reagan's inauguration, the Heritage Foundation issued a report called, "High Frontier: A New National Strategy." This highly influential document advocated expanding military operations and missile defense into space.[8] The tide seemed to be turning in favor of establishing a space command.

The possibility of expanding military operations into space captured the imagination of the new president. In the meantime, the Air Force began preparing an operational home for Space Operations at NORAD/ADCOM headquarters. In January 1982, the Government Accountability Office (GAO) released a report in which it criticized the Air Force for violating the "Single Manager Concept" in its approach to space organization since it dealt with only one branch of service. The GAO recommended a new Combined Space Operations Center in Colorado Springs serve as the "future space command" and recommended withholding funding until DoD "comes up with an overall plan for the military exploitation of space."[9] This report, combined with the Reagan administration's interest in space, spurred the Air Force to further action. Lieutenant General James V. Hartinger, the commander of NORAD, sought to make ADCOM the foundation of a new space command.

The Air Force announced the creation of the Air Force Space Command on June 5, 1982. It became operational on September 1, 1982, and the Air Force activated it with the intention of making it a specified command (meaning it would have official funding and recognition with the DoD and report directly to the JCS rather than to the service Chief).[10] The new command consolidated systems from SAC, ADCOM, and the STC. SAC contributed assets from the Defense Support Program, defense meteorological support program, SPACETRACK, and the Ballistic Missile Early Warning System. The new command's mission revolved around one purpose: space defense. The main component of protecting American military assets in space focused on deploying technology to defeat the Soviet anti-satellite programs. Space Command also had the mission of developing anti-satellite capabilities to counter Soviet satellites. Congress, however, cut funding for these programs. So essentially, the new Space Command had no funding for the mission it had been asked to carry out.[11]

FORMATION OF A UNIFIED SPACE COMMAND

Just 1 month after the Air Force announced the formation of Air Force Space Command, the Reagan administration announced its new national space policy goals: building survivable and enduring space systems, developing an anti-satellite capability, and detecting and reacting to threats to U.S. space systems. This strategic vision would form the foundation for the March 1983 SDI, known as "Star Wars." CJCS General John W. Vessey stated that, "SDI highlighted space's potential as a theater of operations."[12] The administration's vision of an aggressive expansion into space led to the planning for a SPACECOM. The Air Force, which had been the first service to push for a unified space command, took the lead in pushing their existing Space Command forward as the foundation of a unified command. This push elicited a strong reaction from the U.S. Navy and set off 3 years of interservice rivalry that hampered and delayed the formation of a fully unified space command.

On March 7, 1983, *Aviation Weekly & Space Technology* broke the story of the plan to form a joint space command. There was no indication if the new command would be a specified Air Force command or a unified command under the JCS that would involve component organizations from all the services. Space officers in the Air Force were concerned the Army and Navy would not be willing to form the component organizations necessary to have a fully operational unified command. In testimony before the House Armed Services committee, Secretary of the Navy John Lehman and Admiral James D. Watkins opposed the idea of a unified space command. Secretary Lehman claimed, "The whole idea of more and more unified commands in peacetime is a bad idea because it centralizes. Decentralization is clearly the way to gain efficiency in defense matters."[13] Lieutenant General Hartinger, the commander of Air Force Space Command, stated that he would design the unified command to serve all services, saying, "we recognize that spacecraft are indifferent to their customers." Unlike the Navy, the Army was supportive of the move and sent a team to Space Command within a month to get a full briefing on the capabilities, even though it had no command. The plan for the new command would be 50 percent military personnel and 50 percent civilian contractors.[14] Just one month after this article was published, Hartinger forwarded his formal proposal for the organization of a unified space command. He argued that it was necessary in order to implement the recently announced SDI.

The study, *History of the Unified Command Plan 1946-1993,* describes the mission of SPACECOM at its inception as one to "consolidate the mission areas of space control, space support, force application and force enhancement, and exercise operational control over all related systems developed for military application." It went on to state that the Army, Air Force, Navy, and Marine Corps all agreed with the recommendation.[15] But the official agreement over the formation of a unified command papered over a vigorous debate within the Pentagon. On June 19, 1983, *The New York Times* reported that the Navy planned to establish a Navy Space Command on October 1, 1983, consolidating communications, navigation, surveillance, and other space activities within the Navy. In

August, the Navy went through *The Washington Post* to register their official protest:

> The Navy opposes the proposal and plans to create its own space command in Dahlgren, VA, on 1 October. The interservice rivalry on the issue reflects traditional competition for defense dollars as well as differing priorities on how space should be used. . . . Navy officials fear that their need for satellite information would take second place to the Air Force's desire to develop 'Star Wars' weaponry.[16]

According to the *Times*, the Army, on the other hand, was supportive of the new unified command and pledged to support it with Air Defense Assets.[17]

On November 18, 1983, the Pentagon announced that it had decided to form a SPACECOM. *The Washington News* continued the theme of the Navy's resistance to the plan: "The Joint Chiefs of Staff, skirting Navy opposition, have decided to form a unified military command to control defense operations in space."[18] *The Washington Post* provided more detail about internal disagreements when it reported that initially the Joint Chiefs voted 3 to 2 in favor of the plan: the Navy and Marine Corps had opposed the formation of the command. They agreed to forward a unanimous recommendation after the Joint Chiefs agreed to delay the activation date for 2 years, a delay that would give the Navy time to mature the capabilities of its own command. Accordingly, Pentagon leadership saw the Navy's move as an attack on the concept of a unified command, and Navy leadership seemed to hope that, by delaying the formation of the unified command, the very concept of it would be weakened and possibly eliminated.[19]

In February 1984, the JCS issued a planning memorandum for a new space command that would take over the functions of ADCOM. Its primary mission would be to provide tactical warning in combined space operations, control of space, direction of space support activities, and planning for ballistic missile defense. The guidance went into detail to define the relationship between NORAD and SPACECOM due to the many overlapping responsibilities.[20] Organizationally, NORAD would be over SPACECOM, which would be a supporting command to provide NORAD with integrated early warning and assessment information of missile threats. Then NORAD commander General Robert T. Herres recommended combining NORAD with SPACECOM. The CJCS, the Chief of the Air Force, and the Chief of the Army all agreed with this recommendation, but the Navy and Marine Corps Chiefs disagreed.[21] Based on this split recommendation of the JCS, President Reagan ordered Space Command to become fully functional on October 1, 1985. Still, there were service disagreements over management and technical relationships, the role of strategic defense initiatives, who should command, and where the headquarters would be located. The Reagan administration was quite concerned about the Navy's lack of support for the command, so the administration allowed it to maintain relative autonomy within the unified command by keeping the headquarters in Dahlen, VA. The JCS had the opposite problem with the Army, which still needed to define its own space needs in its newly emerging doctrine.[22]

Despite its limited contribution in terms of assets and expertise to the formation of SPACECOM, the Army embraced the formation of the command enthusiastically and began a serious study of how it could make space technology an integral part of the new "Air/Land Battle Doctrine" it was developing. The Army, which had bowed out of space operations after the formation of the U.S. Air Force, took a renewed interest in the available technology. After looking at naval operations, the Army moved to develop technology, namely satellites, that would assist land movements in terms of navigation and communication. In its renewed focus on countering the Soviet threat in Europe, and in relation to its emerging view of space, the Army also developed a concept of "Deep Battle." Critical to the doctrinal concept of Air/Land battle was gaining observation of the enemy's formations in depth before they deployed into battle formation. The Army developed the capability to use Air Force satellites to do imaging reconnaissance in any weather conditions in order to identify concentrations of Soviet armor. The Army moved to develop a space policy and a space office in the Pentagon, and set the groundwork for the future Army Space Command that would eventually become operational in 1988.[23] While the Army saw the need to develop its own space command to put it on an equal footing with the Navy and the Air Force within the SPACECOM structure, it had done so in a way that expertly capitalized on the existing technology, demonstrating the potential of interservice cooperation rather than competition.

On September 23, 1985, DoD activated the U.S. SPACECOM with General Herres as the commander of both NORAD and SPACECOM. Despite budgetary challenges and interservice rivalry, DoD had managed to conceptualize and implement a SPACECOM based around emerging technologies that every service needed. Reagan's SDI provided the necessary impetus to overcome these considerable obstacles. More specific applications came later. For example, when the JCS endorsed the operational requirements proposed for Phase I of the ballistic missile defense system, they placed it under the command of SPACECOM. Clearly, the development of the SDI's broad strategic mandate demanded a coherent and high level integration of complex systems across all services to make it effective.[24] The end of the Cold War decreased the emphasis on space defense and the overall need for space operations, but the groundwork laid before this time helped to revolutionize and unify the way all four services organize, navigate, communicate, and fight.

ENDNOTES - CHAPTER 6

1. Iver Peterson, "U.S. Activates Unit for Space Defense," *The New York Times,* September 24, 1985.

2. L. Parker Temple, *Shades of Gray: National Security and the Evolution of Space Reconnaissance,* Reston, VA: American Institute of Aeronautics and Astronauts, 2005, p. 501.

3. *Ibid.,* p. 502.

4. *Ibid.,* pp. 509-510.

5. *Ibid.,* p. 515.

6. *Ibid.,* p. 518.

7. *Ibid.,* pp. 516-517.

8. *Ibid.,* pp. 520-524.

9. *Ibid.,* pp. 524-525.

10. *Ibid.,* pp. 525-526.

11. *Ibid.,* pp. 526-527.

12. Joint History Office, Chairman of Joint Chiefs of Staff, *The History of the Unified Command Plan*, Washington DC: Government Printing Office, 1995, pp. 95-98.

13. Craig Covalt, "USAF Command Influencing Direction of Programs," *Aviation Week & Space Technology*, Vol. 7, March 1983, p. 44.

14. *Ibid.*, p. 45.

15. *The History of the Unified Command Plan*, p. 96.

16. Fred Hiatt, "Military Considers Unified Space Command," *The Washington Post*, August 27, 1983.

17. Richard Halloran, "Air Force Seeking Joint Space Unit," *The New York Times*, June 19, 1983.

18. Richard C. Gross, "Joint Chief Recommends Unified Space Command," United Press International, *Washington News*, November 18, 1983.

19. Fred Hiatt, "Pentagon Proposal; Joint Space Command Sought," *The Washington Post*, November 18, 1983.

20. Temple, p. 536.

21. *The History of the Unified Command Plan*, p. 96.

22. "Joint Staff to Shape Space Command," *Aviation Week & Space Technology*, Vol. 10, December 1984, p. 21.

23. "Army Reviewing Space Program Efforts," *Aviation Week & Space Technology*, Vol. 28, January 1985, p. 21.

24. *The History of the Unified Command Plan*, p. 97.

PART II:

SUB-UNIFIED COMMANDS AND ORGANIZATIONS

CHAPTER 7

U.S. CYBER COMMAND'S ROAD TO FULL OPERATIONAL CAPABILITY

Michael Warner

INTRODUCTION

U.S. Cyber Command (CYBERCOM) achieved full operational capability in October 2010 as a sub-unified command under U.S. Strategic Command. Its course to this status took several turns due to a number of factors related mostly to the novelty of the cyber domain, which left considerable uncertainty in the minds of decisionmakers at several levels in the Department of Defense (DoD). What ultimately prevailed was the strong support of the Secretary and the conviction among senior defense leaders—even as they debated the particulars—that the nation needed something done swiftly to defend military networks. The main lesson of U.S. Cyber Command's accomplishment thus would seem to relate to the centrality of national-level policy concerns even in military matters. Secondary lessons include the importance of staff coordination and the staff's command of information vital to decisionmaking processes.

CYBERCOM's attainment of full operational capability (FOC) status took roughly 2 years from the time Secretary of Defense Robert M. Gates set the process in motion. In many ways, the process toward FOC typified the establishment of a major organization in the DoD, but in other respects, the novelty of the cyber domain—in which every Service, combatant command, and agency operates and even "fights"—added unforeseen complexity to decisionmakers' roles. Indeed, nearly every senior leader in the Department had some equity that would be affected by the work of the new CYBERCOM, and many of those leaders had advice for the principals making the key decisions about it.

An examination of CYBERCOM's progress to FOC thus has to be more than a chronicle of the key events and relevant leadership actions. The formation of a major new defense organization in a new battlespace is automatically a primer in organizational change. This chapter surveys the events leading to FOC and reflects on their significance by drawing upon the documentation assembled by the CYBERCOM team that managed the process, supplemented not only by the memories of the team members but also by research in Command records. It is by no means definitive, but its accuracy and timeliness should complement the breadth and depth of research that will be possible in the future.

ANTECEDENTS

The Information Revolution has empowered people and institutions to work more efficiently and take advantage of unprecedented opportunities. At the same time, however, the networking of the world's information systems in "cyberspace" has opened new fields for criminality and coercion, and tied the security of private individuals to that of enterprises and nations in unforeseen ways. The importance of cyberspace to national security became a pressing concern after the end of the Cold War. Such concerns increased dramatically as exercises like "Eligible Receiver 97" demonstrated network vulnerabilities and, as American officials discovered with the Moonlight Maze incident in 1998, that foreign entities had been probing sensitive U.S. military networks.[1] The Joint Chiefs of Staff (JCS) in their 2004 *National Military Strategy* declared cyberspace a domain (like air, land, sea, and space) in which the United States must maintain its ability to operate.

The DoD and the Armed Services responded to these evolving challenges through a variety of organizational initiatives. The first of these was the Joint Task Force-Computer Network Defense (JTF-CND), a small organization chartered by the Secretary of Defense and reporting directly to him. JTF-CND operated in conjunction with the Department's de facto Internet service provider, the Defense Information Systems Agency (DISA), and attained initial operating capability on December 1, 1998.[2] President Bill Clinton under Unified Command Plan 1999 soon assigned JTF-CND to U.S. Space Command (SPACECOM). The offensive and defensive cyber missions came together under the same organization in 2000, when SPACECOM formally took over the DoD computer network attack planning. As a result, JTF-CND was re-designated the Joint Task Force-Computer Network Operations (JTF-CNO) in April 2001. When SPACECOM was dissolved and its functions merged into the reorganized U.S. Strategic Command (USSTRATCOM) on October 1, 2002, JTF-CNO had 122 positions and a $26 million budget. Its new mission, under Strategic Command and with the geographic combatant commands, was to:

> coordinate and direct the defense of DoD computer systems and networks; [and] coordinate and, when directed, conduct computer network attack in support of combatant commanders' and national objectives.[3]

JTF-CNO was headquartered in Arlington, VA, with DISA's Global Network Operations and Security Center (GNOSC), and had a 24-hour watch floor there.

In 2002, the transfer of Defense-wide computer network operations responsibilities to USSTRATCOM occurred as discussions in the Department over these roles were increasing. USSTRATCOM soon approved the Joint Concept of Operations for Global Information Grid Network Operations. In June 2004, Secretary of Defense Donald Rumsfeld added the final step in this transformation by authorizing the creation of the Joint Task Force-Global Network Operations (JTF-GNO), with the three-star Director of DISA dual-hatted as its Commander (and as USSTRATCOM's Deputy Commander for Network Operations and Defense). The next year, Strategic Command's General James Cartwright (USMC) completed the task of rearranging USSTRATCOM by creating a series of joint functional component commands to perform the Command's various missions. The new Joint Functional Component Command for Network Warfare (JFCC-NW) would be commanded by the Director of the National Security Agency (NSA) and take on the offensive side of the now-defunct JTF-CNO's responsibilities.

When USSTRATCOM finished its reorganization, DoD had assembled a complicated arrangement of cyber capabilities and organizations. DoD also provided information technology services Department-wide via DISA; used NSA for cyber intelligence and information assurance; and administered some policy and oversight functions in the office of the Assistant Secretary of Defense of Networks and Information Integration (who was also DoD's Chief Information Officer). USSTRATCOM grouped its military cyber capabilities (both offensive and defensive) in two organizations: JFCC-NW was paired with NSA, and JTF-GNO with DISA. Those two partnerships gave the offensive and defensive operators, respectively, access to subject matter expertise, but their bifurcation also meant that they talked less to one another than they had under the old JTF-CNO. Each Service had its own cyber component, moreover, to manage its own networks. This congeries of capabilities fully satisfied no one, and within 2 years a high-level effort to revise it was underway.

INITIAL DECISIONS IN 2008

In early-2008, Secretary of Defense Robert Gates wondered about better ways to organize the DoD's cyber functions, setting in motion studies of alternatives to the current arrangement. Indeed, the possibility of a "Cyber Command" had been discussed that February by General Kevin P. Chilton, the new Commander of USSTRATCOM, and senior officials from the Pentagon, Washington, DC, NSA, and the Office of the Director of National Intelligence. This preliminary work led to the Secretary's direction in May 2008 to task a Departmental-level review of cyber roles and missions, to be conducted by the Quadrennial Roles and Missions Review's Cyber Team. The team considered reorganization schemes that summer under the supervision of Principal Deputy Undersecretary of Defense (Policy) Christopher "Ryan" Henry and USSTRATCOM's Deputy Commander, Vice Admiral Carl V. Mauney (USN). This effort was among the earliest to contemplate the creation of a "Cyber Command," and it revived the notion that the new entity should oversee both the offensive and defensive facets of cyber operations. Another study group, led by a former U.S. Air Force Chief of Staff, General Larry Welch, evaluated the issues for the Joint Chiefs under the auspices of the Institute for Defense Analyses (Welch was that organization's president). In sum, it appears that a consensus had emerged that the current division of labor between DoD cyber security and network attack organizations was sub-optimal and needed to be changed sooner rather than later. Secretary Gates heard the briefs, and on October 2, 2008, he "indicated that a four-star sub-unified Command under USSTRATCOM should be DoD's organizational endstate for cyber C2 [command and control]."[4]

At this point, Secretary Gates declined to decide the new entity's ultimate configuration and instead, on November 12, 2008 realigned the existing organizations. Citing "a pressing need to ensure a single command structure is empowered to plan, execute, and integrate the full range of military cyberspace missions," he directed USSTRATCOM, effective immediately, to "place [JTF-GNO] under operational control of Commander [JFCC-NW]."[5] This added a new job to the duties of Lieutenant General Keith B. Alexander (United States Army), who was already serving as both Director of NSA and Commander of JFCC-NW. More important, it meant that both the offensive and defensive components of DoD cyber capabilities would, for the first time, operate in close proximity to the nation's signals intelligence system.

Several events factored in the Secretary's thinking and the timing of his order. In particular, NSA had played a key role in detecting the presence of foreign intelligence malware in DoD classified networks in October 2008, and was helping DoD organizations neutralize the infection in an operation named BUCKSHOT YANKEE.[6] Additionally, Secretary Gates was, at this point, reasonably certain he would be asked to stay on under the incoming administration of President-elect Barack Obama, which would allow him to implement broader changes he was directing in DoD cyberspace organizations.

FORMING A COMMAND, JANUARY 2009-MAY 2010

President Obama took office on January 20, 2009, and, by coincidence or not, discussions over implementing the Secretary of Defense's order around that time took a decisive turn. The previous month, a blue-ribbon panel convened to advise Secretary Gates on managing the nuclear weapons stockpile had concluded that USSTRATCOM had too many missions, and publicly recommended that the Command's responsibilities be narrowed to nuclear matters only (leaving cyberspace and other missions to other DoD organizations).[7] In March, USSTRATCOM assembled a team of planners to work with NSA and JFCC-NW experts at Fort Meade, MD, to develop a commanders' estimate, which Alexander could use to explain to Chilton how he planned to exercise the operational control of JTF-GNO, granted him the previous November. The estimate's scope was expanded in April, however, to encompass options for a new Cyber Command, shortly before rumors of a new military command hit the news media.[8] Alexander briefed Chilton on May 1 on the progress toward the commander's estimate.

A few days later, Alexander explained to the House Armed Services Committee in a public session that the replacement of analog technologies by digital networks meant the world was now linked in "the same network." The U.S. military had seized opportunities resulting from this development but was not yet addressing the accompanying risks; indeed, in Alexander's view, the current approach to cyber security "does not work." Hinting at the DoD impending decision, he added:

> we're looking at the steps of what we have to put together in the sub-unified command as an option, or in a Joint Functional Component Command—how will we put these capabilities together to ensure our networks are secure and provide us freedom of maneuver in cyberspace.[9]

Secretary Gates gave his answer on June 23, 2009. "Effective immediately," he directed USSTRATCOM "to establish a subordinate unified command designated as U.S. Cyber Command (USCYBERCOM)." JFCC-NW and JTF-GNO would be dismantled and their personnel reassigned to USCYBERCOM, which the Secretary "preferred" to see based at Fort Meade with NSA. The Joint Chiefs of Staff were to issue a planning order to USSTRATCOM to develop an implementation plan, and initial operating capability was to be reached by October 2009, with full operating capability following in October 2010. USCYBERCOM was also authorized direct liaison privileges with the geographic combatant commands.[10]

USSTRATCOM responded smartly to the Secretary's direction. The commander's estimate team had already been re-chartered as the "Implementation Planning Team" 2 weeks earlier. Talks between senior officers from USSTRATCOM, NSA, JFCC-NW, JTF-GNO, and DISA set the stage for the Implementation Planning Team's work. Meeting at NSA, the team started drafting an Implementation Plan and created a "cyber story board" to explain the emerging concepts. That brief served as the basis for briefings delivered across Washington and the military in ensuing months. Meanwhile, Chilton sent the finished Plan to the Chairman on September 1; it listed 13 required tasks for reaching initial operational capacity (IOC) but did not set hard criteria for determining FOC. Instead, the Plan included several dozen tasks of varying importance and specificity to complete by October 1, 2010, in its larger matrix of actions for attention between 2009 and 2011.[11] At FOC, the Plan's "Commander's Intent" was that:

> USSTRATCOM [Unified Command Plan] authorities and planning responsibilities related to cyberspace will have been transferred to CDRUSCYBERCOM [Commander, USCYBERCOM], and USCYBERCOM's capacity and capabilities for cyberspace operations will have matured to a point where it can plan, synchronize, and execute cyberspace operations as a supported or supporting command.[12]

The new organization soon began to grow, building on existing JTF-GNO and JFCC-NW manpower. On October 16, 2009, President Obama nominated Alexander to be the first Commander of USCYBERCOM. A couple weeks earlier (on October 5), JFCC-NW and JTF-GNO had begun to merge their staffs and operational centers into a consolidated staff. It in turn began hiring senior officials to head its "J-Code" directorates.[13] Many of the functions of the JFCC-NW Deputy Commander now went to the new chief of staff, Major General David N. Senty (United States Air Force Reserve), to manage for the consolidated staff.

LOW-HANGING FRUIT, MARCH TO AUGUST 2010

The Pentagon had expected confirmation hearings for Alexander before the end of the year. For a number of reasons, however, the confirmation was delayed.[14] Alexander testified before the Senate Armed Services Committee on April 15, 2010, and 3 weeks later, the Senate approved both his nomination to head the new USCYBERCOM and his promotion to general.[15] With this step taken, on May 21, the Secretary of Defense presided at a promotion ceremony in NSA headquarters, deactivating JFCC-NW, and declaring that USCYBERCOM had achieved IOC.

The new Command had to have a way of measuring progress toward FOC. In April, Senty began tracking a series of metrics based on the Implementation Plan (I-Plan) and a dozen commander's priorities that his staff had recently crafted for then-Lieutenant General Alexander.[16] On July 27, Senty's staff, which had helped draft the I-Plan, requested Command staffs to provide weekly details of progress toward FOC and a re-validated list of milestones.[17] The resulting "Strategy to Tasks Status Update" brief sorted dozens of I-Plan actions according to the commander's priorities into 23 tasks and placed them in a matrix that would be the main device for tracking progress. The work paid off that summer and fall, when the staff's matrix repeatedly won praise for the situational awareness it provided to senior leaders.

Creating situational awareness in cyberspace was also vital for the new Command. On March 5, 2010, the consolidated staff merged watch personnel and expertise (mostly from JTF-GNO and JFCC-NW) in a combined Joint Operations Center (JOC) at Fort Meade. Control of USCYBERCOM operations from the combined JOC began on May 17, along with new procedures designed to "operationalize" DoD information networks. By August, the JOC was functioning well enough to continue JTF-GNO's watch function.

Another hurdle was the move and assimilation of JTF-GNO, the planning for which had begun by August 2009. In Fiscal Year 2010, JTF-GNO was authorized 66 military and 138 civilian personnel, some of whom would return to DISA, their parent organization. Many of those who chose to go with USCYBERCOM needed upgraded security clearances before they could effectively support the Command's mission at Fort Meade. The upgraded clearances all sat with the DISA in Arlington, VA (DISA's Director, Lieutenant General Carroll Pollett, also served as JTF-GNO's Commander). That in itself added another complicating factor, as DISA had been slated by the Base Realignment and Closing (BRAC) process in 2005 to move to a new headquarters building at Fort Meade in 2011. Thus, DISA was in the midst of planning its own move north. All these factors came together to delay JTF-GNO's transition. USCYBERCOM had assumed JTF-GNO's command and control functions by early-June, but the full transition of personnel and databases that had been slated to occur on June 30 had to be pushed back 2 months. JTF-GNO was formally disestablished at a ceremony at DISA on September 7.

Finally, on August 5, the Senate confirmed the nomination of Major General Robert E. Schmidle (USMC) to be the first Deputy Commander, USCYBERCOM; he was promoted to lieutenant general 4 days later and reported for duty on August 10. These moves set in place many of the personnel and structural issues that came with setting up a new command, and the new command had effectively accomplished several organizational tasks that demonstrated progress toward FOC, but challenges still remained.

TOUGH ISSUES, JUNE TO OCTOBER 2010

Several issues seemed likely to persist even after the declaration of FOC. As Senty's aide explained, these issues amounted to "building capability and capacity in Service cyber forces, and gaining the requisite authorities and fully resourcing the Command."[18] Each presented an interlocking series of complications for every decisionmaker who approached it. Gates himself introduced another problem set in August. The issues collectively prompted high-level debates over the wisdom of declaring the Command to be in FOC status later rather than sooner.

The first set of challenges revolved around questions of authority. What authorities would USCYBERCOM possess? As a sub-unified command, it operated under the authorities delegated to it by its institutional parent, USSTRATCOM, which in turn were derived from the Unified Command Plan approved by the outgoing President George W. Bush in December 2008. That same month, Gates had directed USSTRATCOM to draft a global campaign plan to secure, defend, and operate DoD information systems. USSTRATCOM had responded with Operation GLADIATOR PHOENIX, delivering a draft of its execute order to the Joint Staff in May 2009. Staffing and coordinating the order in the Department began promptly, but with the change in direction dictated by Gates' announcement of his intent to create a cyber command that June and the delay in the Commander's confirmation, it was not completed until after the Command reached IOC a year later. Chilton briefed Gates on GLADIATOR PHOENIX in June 2010, and Gates approved it on February 11, 2011.

USCYBERCOM also had to determine how it would exercise command and control over the Service cyber components that were to be assigned to it. USCYBERCOM also had to plan how it would integrate its operations with those of the geographic combatant commands. The Joint Chiefs discussed these challenges in August 2010, directing the Command to run a series of tabletop exercises to identify the relevant issues. Schmidle ran the first of these at Fort Meade in October. The event helped to demonstrate that the Command was assuming its responsibility to advance the debates over lines-of-authority in the new cyber domain.

As questions about USCYBERCOM's authority were ironed out, questions about resourcing emerged. The new Command's leaders waited months to learn which Service units the Pentagon would assign to USCYBERCOM. The Services had begun reorganizing their cyber capabilities in late-2009, with the idea of creating headquarters units (in addition to those already assigned to USSTRATCOM) that would function with the proposed USCYBERCOM. Over the next few months, the Services created the Army Cyber Command; Marine Forces Cyber Command; Fleet Cyber Command/U.S. Tenth Fleet; and Air Force Cyber Command/24th Air Force. As USCYBERCOM neared its FOC date, however, these forces remained in an institutional limbo, not yet assigned to any command. Gates approved the new Assignment Tables for all the unified commands only in December 2010—after FOC—and USSTRATCOM delegated operational control of various Service cyber units and their headquarters to USCYBERCOM a few days later.

A third set of questions about efficiencies also emerged close to the FOC target date. On August 9, 2010, Gates added another consideration for decision-makers at USSTRATCOM and USCYBERCOM. Speaking at a Pentagon press conference, he announced broad budget cuts across the DoD; defense agencies and unified commands in particular were to hold their future personnel totals to Fiscal Year (FY) 2010 levels. USCYBERCOM had not been scheduled to receive real increases in manning until FY11. Its combined JFCC-NW and JTF-GNO numbers totaled just over 500 FY10 billets, vice the 900-plus it had been projected to have in FY11 to perform its significantly expanded mission. The Command formally appealed for an exception in September, and several weeks later Deputy Secretary of Defense William Lynn granted the request.

At the same press conference on August 9, Secretary Gates also announced his intention to change the way the Department organized itself to administer its information networks. The Secretary stated a desire to shed the Assistant Secretary of Defense for Networks and Information Integration (ASD/NII) position and to divide its functions (along with some of those of the Joint Staff's J6) between DISA, the DoD Chief Information Officer, and possibly USCYBERCOM as well. Behind the scenes, moreover, another move was afoot. Gates quietly asked for options for increasing DoD reliance on a network architecture derived from a "cloud computing" proposal. The possibility of expanding USCYBERCOM size and mission was very much on the minds of senior defense officials as the date set for FOC drew near.

With Gates' October deadline for FOC approaching, Alexander noted USCYBERCOM accomplishments on the road-to-FOC tasks listed by the Secretary, noting the remaining challenges (IT efficiencies, manpower, and personnel), and recommended the Secretary approve the declaration of FOC. Although concerns persisted over the risks created by those gaps, Chairman of the Joint Chiefs of Staff Admiral Michael Mullen (USN) and Vice Chairman General James Cartwright (USMC) urged the Secretary to declare USCYBERCOM to be in FOC status. On Sunday, October 31, 2010, Lynn approved FOC for USCYBERCOM. His statement noted that the Command had accomplished Gates' five critical tasks from the previous May, and ordered USSTRATCOM to articulate requirements for personnel, authorities, and information technologies efficiencies.[19] Thus, in a sense, USCYBERCOM's real work was just beginning.

CONCLUSION

The creation of USCYBERCOM marked the culmination of more than a decade's worth of institutional change. DoD defensive and offensive capabilities were now firmly linked, and, moreover, tied closely, with the nation's cryptologic system and premier information assurance entity, the NSA. That interlocking set of authorities, personnel, and organizations would also be better able to partner with both the geographic combatant commands and other U.S. Government agencies to defend the nation in cyberspace and ensure its freedom to maneuver in this new and challenging domain.

In organizational terms, USCYBERCOM's stand up represented an enormous amount of work performed at a fast pace. Despite a compressed schedule, the consolidated staff at USCYBERCOM and the legacy organizations it subsumed were able to accomplish a great deal by October 2010. They established a Joint Operations Center at Fort Meade, and disestablished USSTRATCOM's Joint Functional Component Command for Network Warfare as well as its Joint Task Force for Global Network Operations. The latter task took considerable planning and effort because JTF-GNO's activities and workforce had to be moved from Northern Virginia to Fort Meade while leaving the daily functioning of DoD information networks unimpaired. The consolidated staff fashioned effective command and control of cyber forces in the Services and reinforced a good working relationship with the DISA. It installed liaison officers at the combatant commands and cyber support elements as well, and deployed expeditionary teams to support operations in Iraq and Afghanistan. It also made progress in support of operational planning by the combatant commanders and in building processes for them to issue requirements for cyber support. In addition, the Consolidated Staff completed actions or made progress on a number of other matters, and accomplished all of this relatively seamlessly, keeping DoD operations secure while making the transition transparent to users of its information systems.[20]

Three important issues remained unresolved at USCYBERCOM's attainment of FOC. First, DoD had a shortfall of assigned cyber force capacity to plan, operate, and defend its networks and ensure freedom to access and maneuver in cyberspace. Second, the Command inherited authorities from predecessor organizations that seemed sufficient to defend DoD networks, but insufficient to protect the U.S. Government's networks or those associated with critical infrastructure in ways that the evolving cyber threat seemed to require. Thus, there was a respectful airing of views in 2010 over the levels of risk associated with various options for pushing forward with a declaration of FOC for a new organization in a new domain. What drove the decision in the end was the leadership and support of the Secretary, as well as the conviction among senior leaders. Even as they debated the particulars, they agreed with one another that, because the nation needed something done swiftly to defend its military networks, it was riskier to hold USCYBERCOM in an indeterminate status than to advance its formation despite the lack of final resolution for these tough issues.

The process by which USCYBERCOM reached FOC was unique because cyberspace is a unique domain. Nonetheless, the events are worth recounting and patterns noticed because they have relevance for organizational change in DoD and for other sorts of organizations adapting to work in cyberspace. In this vein, there are several observations that might have more broadly applicable significance across DoD, particularly in regard to the attainment and declaration of FOC for a new command.

First and foremost, an FOC declaration for a major command entity is inherently a policy (and perhaps political) judgment. It broadcasts as U.S. policy the DoD belief that it could one day have to fight in a certain place or in a certain manner. Therefore, no determination of FOC can ever be entirely military in nature—and thus it will be driven by considerations partly outside of "objective" criteria and metrics. Similarly, IOC and FOC are the Secretary's to set and determine and declare. It is difficult to know in advance just how the world, the threat, and DoD will look as FOC nears. His vote on whether an entity is "ready" is the only one that counts. All other DoD actors in the process serve in an advisory capacity.

The policy implications notwithstanding, setting criteria for an FOC determination is important, as everyone concerned has to live with the results of a declaration when it comes. Criteria should be set early and well—and not chosen in any sort of ad hoc manner. Their meaning and relative centrality for FOC need to be understood, and they should not be changed as the process unfolds (they are either met, or unmet). When new items or tasks are added or obsolete ones removed from a list of FOC criteria, such an amendment needs to be executed with copious documentation and justification—in short, transparency. The initial standards for FOC should also designate the entity authorized to make such amendments and explain the process for doing so.

Organizationally, "Stoplight charts" or other metrics for criteria impose salutary discipline on the analysis of progress toward FOC. They also help seniors and staff to coordinate their perceptions and their actions. Obviously they are only a tool, however, and should not come to represent an end in themselves. Finally, staffs need to use those tools to coordinate with one another. IOC and FOC by definition involve a new staff emerging from an existing one. Both staffs must be synchronized. This is doubly tough to accomplish when the staffs are geographically separate—which only increases its importance.

ENDNOTES - CHAPTER 7

1. James Adams, "Virtual Defense," *Foreign Affairs*, May/June 2001.

2. Timothy Madden, *A Legacy of Excellence*, Arlington, VA: Joint Task Force-Global Network Operations, 2010, p. 6.

3. U.S. Strategic Command Public Affairs press release, February 2003, available from *www.iwar.org.uk/iwar/resources/JIOC/computer-network-operations.htm*. See also Ellen Nakashima, "Warriors in the Battle for Cyberspace," *The Washington Post*, September 25, 2010.

4. Joint Staff, J-5 Cyber Division, "Proposed SecDef Memorandum to Establish a Subordinated Unified Command for Cyber under USSTRATCOM," Joint Staff Action Processing form for J-5A 30217-09, April 20, 2009.

5. Secretary of Defense to Service, Command, and agency heads, "Command and Control for Military Cyberspace Missions," Washington, DC: DoD, November 12, 2008. The Secretary also stipulated that DISA's Director "will continue to serve as Commander of JTF-GNO and will remain responsible providing the JTF-GNO network and information assurance technical assistance as required."

6. Ellen Nakashima, "Cyber Intruder Sparks Response, Debate," *The Washington Post*, December 8, 2011.

7. James Schlesinger *et al.*, "Report to the Secretary of Defense on DoD Nuclear Weapons Management," December 2008, p. 54.

8. Staff e-mail to planning team invitees, "Cyber Mission Planning Team Off-Site," April 9, 2009; Siobhan Gorman and Yochi J. Dreazen, "New Military Command to Focus on Cybersecurity," *Wall Street Journal* online, April 21, 2009.

9. Testimony before the House Committee on Armed Services, Subcommittee on Terrorism, Unconventional Threats and Capabilities, May 5, 2009, pp. 7-8.

10. Robert M. Gates, Secretary of Defense, "Establishment of a Subordinate Unified U.S. Cyber Command under U.S. Strategic Command for Military Cyberspace Operations," June 23, 2009.

11. "Initial operational capability" (IOC) is a term borrowed from the defense acquisition world to denote attainment of a significant ability to wield a new weapon or system as deployed in field conditions. There is no consensus on how long IOC should

take for a large DoD organization. The closest historical parallel to the establishment of USCYBERCOM, both institutionally and chronologically, was the establishment of U.S. Africa Command, which was announced in February 2007, reached IOC on October 1, 2007, and attained FOC on October 1, 2008.

12. USSTRATCOM, "United States Cyber Command Implementation Plan," September 1, 2009, p. 43.

13. For example, Major General David Lacquement (United States Army) came aboard in September as J3, and Major General Suzanne M. Vautrinot (United States Air Force), JFCC-NW's Deputy Commander, wore two hats as the Staff's J5.

14. The Senate Armed Services Committee scheduled the hearing after its staffers were briefed on plans for USCYBERCOM. Briefings of key members and staffers began in November 2009.

15. The Senate approved the nomination by unanimous consent on May 7, 2010.

16. This was the result of a series of meetings of the Consolidated Staff from December 2009 to February 2010, culminated by a Staff brief of Vice Admiral Mauney at Fort Meade in March 2010.

17. USCYBERCOM Chief of Staff's office, e-mail to multiple recipients, "Report Card on CDR's [commander's] Priorities—New Weekly Reporting—Due 4 Aug," July 27, 2010.

18. USCYBERCOM J0, "USCYBERCOM's Road to Full Operational Capability [FOC])," Information Paper, August 11, 2010 (U/FOUO).

19. Deputy Secretary of Defense to Commander, USSTRATCOM, "Full Operational Capability (FOC) of US Cyber Command (USCYBERCOM)," October 31, 2010,

20. Keith B. Alexander, "Building a New Command in Cyberspace," *Strategic Studies Quarterly*, Vol. 5, No. 2, Summer 2011, pp. 3-4.

CHAPTER 8

CORDS IN CHARGE: ORGANIZING FOR PACIFICATION SUPPORT IN THE VIETNAM WAR

Gregory A. Daddis

INTRODUCTION

Most American estimates of South Vietnam in early-1965 painted a grim picture. Political instability wracked the Saigon government. National Liberation Force (NLF) insurgents posed both a political and military threat in the countryside and increasingly showed a willingness to confront South Vietnamese Army (ARVN) units in battle. Political subversion, assassination of government officials, and attacks on infrastructure continued at an alarming rate. Equally grave, American intelligence analysts picked up indications of regular army units from North Vietnam infiltrating into the south.[1] That spring, U.S. Military Assistance Command, Vietnam (MACV) commander General William C. Westmoreland, who believed the American advisory effort had done all it could to support the teetering Saigon government, requested the introduction of ground combat troops. On June 26, 1965, Washington officials gave the general permission "to commit U.S. ground forces anywhere in the country when, in his judgment, they were needed to strengthen South Vietnamese forces." The number of American forces rose precipitously. At the beginning of 1965, 23,000 U.S. troops were in Vietnam. One year later, the number soared to 184,000 troops.[2]

The troop escalation mirrored the growing concerns of Westmoreland and his officers. Their analysis of the operational environment, and the nature of the threat, led them to believe the enemy was doing all it could "to destroy the [South Vietnamese] government's troops and eliminate all vestiges of government control."[3] In fact, this assessment struck close to the mark. Responding to U.S. intervention, planners in Hanoi, North Vietnam, became convinced, albeit not without disagreement, that only by escalating the struggle inside South Vietnam could they eventually defeat the "American aggressors." In March 1965, Hanoi's Party Central Committee decided to "mobilize the soldiers and civilians of the entire nation to strengthen our offensive posture and to attack the enemy." Especially important, however, is that this strategic offensive encompassed more than just military escalation. Communist leaders in South Vietnam spent considerable energy on "a wide-ranging political campaign throughout the armed forces and the civilian population" to minimize the effects of American intervention.[4] This combined political-military campaign was aimed at orchestrating the efforts of regular and insurgent forces, as well as coordinating the activities of political cadres and military units.

The point was not lost on the MACV commander. As early as mid-1965, Westmoreland realized that simply destroying enemy forces would not secure victory in South Vietnam. Even before the November 1965 clash between American and North Vietnamese regulars in the Ia Drang Valley, MACV's chief maintained that the "war in Vietnam is a political as well as a military war."[5] The problem for the Americans, however, centered on building an organization that could manage the disparate tasks of a strategy built on both political and military objectives. "Probably the fundamental issue is the question of the coordination of mission activities in Saigon," Westmoreland opined in early-1966.

> It is abundantly clear that all political, military, economic, and security (police) programs must be completely integrated in order to attain any kind of success in a country which has been greatly weakened by prolonged conflict and is under increasing pressure by large military and subversive forces.[6]

Despite this perceptive assessment, it wasn't until 1967 that the American military command in Vietnam created an organization to coordinate the nonmilitary aspects of Westmoreland's strategy. In the process, American leaders, both in Washington and Saigon, relied on historical analogies to argue for a reorganization of the pacification effort in South Vietnam. This chapter examines the process such leaders underwent to create an organization for managing pacification, the wartime effort to mobilize popular support across the South Vietnamese countryside for the Saigon government. Historical antecedents appeared to offer clear guidance on organizing for success in an unconventional environment.[7] Only through a unified effort with fully coordinated civil and military functions could one hope to defeat an insurgency. While the creation of a pacification organization went far in coordinating such functions, in the end American attempts to influence popular support among the South Vietnamese countryside proved inadequate. Reorganization could remedy neither ingrained problems nor inconsistencies within a society so long subjected to war.

PACIFICATION – A FITFUL START

Unsurprisingly, the South Vietnamese themselves realized that the insurgent threat required more than just a military approach. In March 1964, the Saigon government announced its *Chien Thang* (Victory) National Pacification Plan, which integrated economic, social, and governmental reform efforts.[8] The program, supported by American advisors, drew heavily from the British experience in Malaya. The successful counterinsurgency against the Malayan Communist Party in the 1950s seemed to offer a model for success in counter-revolutionary warfare, especially in the wake of the French defeat in Indochina. British specialists such as Sir Robert Thompson argued forcefully for "a proper balance between the military and civil effort, with complete coordination in all fields."[9] While the *Chien Thang* program faltered due to governmental instability – coups had wracked Saigon since President Ngo Dinh Diem's assassination in 1963 – it nonetheless established a foundation for future efforts in synchronizing political and military functions.

In supporting *Chien Thang,* American advisors found historical analogies like Malaya all too alluring. The British experience not only informed early pacification efforts in Vietnam, but also validated U.S. Army counterinsurgency doctrine, which counseled commanders that "police operations, civic action, and combat operations against the guerrilla force" should all be "conducted concurrently."[10] The problem, however, centered less on creditable intentions than on implementation in the field. The increasing American commitment in Vietnam brought confusion over who was responsible for the growing number of military and civilian agencies and personnel operating within the war-torn country: MACV, the Agency for International Development (USAID), the United States Information Agency (USIA), and the Central Intelligence Agency (CIA). As one army colonel noted, "everybody is wandering around without any clear-cut direction and management." Both historical precedence and army doctrine clearly advocated a joint effort against the growing insurgency in South Vietnam, yet all levels of the organization lacked such coordination.[11]

By early-1966, Washington officials could no longer ignore the inadequacy of governmental coordination in Vietnam. At the February Honolulu Conference, President Lyndon B. Johnson met with South Vietnamese leaders to discuss the rising importance of pacification. Johnson's Secretary of Defense, Robert McNamara, concurrently offered a grim assessment, noting that pacification was a "basic disappointment." McNamara further concluded that "part of the problem undoubtedly lies in bad management on the American . . . side."[12] This pressure from civilian leadership proved an important, if not essential, element of organizational change. The President's interest in what was increasingly being called the "other war" left both civilian and military war managers little choice but to embrace the task of organizing for pacification support. Westmoreland, in attendance at the Honolulu Conference, returned to Saigon and dutifully began placing increased emphasis on pacification and revolutionary development.[13]

In fact, many uniformed officers were reaching similar conclusions as Johnson and McNamara. In March 1966, the Army staff published a report on the war titled "A Program for the Pacification and Long-Term Development of South Vietnam." PROVN for short, the report charged that "interagency competition" within the American mission in Vietnam was a major obstacle hindering the achievement of U.S. objectives.[14] Westmoreland followed suit, placing command emphasis on revolutionary development and civic action programs and noting in his strategic guidance for 1967 that the pacification effort should "properly dovetail the military and civil programs." Even officers returning from South Vietnam were advocating change. One lieutenant colonel, writing his student essay at the U.S. Army War College in mid-1966, highlighted the need for an "integrated strategy" that synchronized the various military, economic, social, psychological, and political aspects of the war in Vietnam.[15]

This confluence of external and internal stimuli for organizational change provoked a still-reluctant U.S. Embassy in Saigon to create the Office of Civil Operations (OCO) in November 1966. Although embassy officials feared that OCO would lead to a military takeover of civilian programs—Westmoreland supported MACV serving as the "single manager" for pacification—the new office directly improved supervision of the pacification effort's civil side.[16] OCO unified interagency direction and created a pacification chain of command from Saigon to the countryside's districts and provinces. Senior officials working on pacification, from the CIA to USAID, now worked together in a central location, which facilitated planning and coordination. The office consisted of six program divisions run by nearly 1,000 American civilians operating on a budget of $128 million. OCO now managed refugee programs, revolutionary development cadre training, psychological operations, and public safety planning. The military side of pacification, however, remained outside of OCO's purview. Thus, while the office served as the first full step towards a new pacification organization, the "other war" remained separated from military operations being conducted by MACV, including those related to pacification. Less than 6 months later, American officials, citing a visible lack of improvement in the field, dismantled OCO and incorporated it into a new organization.[17]

CIVIL OPERATIONS AND REVOLUTIONARY DEVELOPMENT SUPPORT (CORDS) IN CHARGE

Despite its size, OCO simply did not have the resources to implement the programs for which it provided oversight. Westmoreland's strategic concept for 1967, which included more than just attrition of enemy forces, left OCO increasingly unable to cope with the coordination of civil and military efforts. Westmoreland recalled that as:

> the American military effort expanded, so did the programs managed by [US]AID, CIA, and USIA, so that in time all agencies were competing for resources and scarce South Vietnamese manpower.[18]

The problem simply was too large and complex for OCO to handle alone, and once more historical analogies contributed to reorganizing for pacification support. Again, the British experience in Malaya seemed to corroborate American claims that population security necessarily preceded pacification of the countryside. (The Army's officer corps largely viewed counterinsurgency as a sequential process in which security served as the prerequisite for governmental and social reform.) If OCO did not have the resources or capabilities to attain this security, it increasingly became clear that only one component of the U.S. mission in Vietnam did have such means.[19]

On May 9, 1967, President Johnson charged MACV with responsibility as the "single manager" of pacification in South Vietnam. The President appointed Robert W. Komer, a long-time CIA analyst and National Security Council staff member, as Westmoreland's deputy for pacification. As Johnson declared, this "new organizational arrangement represents an unprecedented melding of civil and military responsibilities to meet the overriding requirements of Viet Nam."[20] Holding ambassadorial rank, Komer assumed control of the newly-created office of CORDs and reported directly to Westmoreland (see Figure 8-1).[21] The new CORDS chief was not an advisor or coordinator but rather held broad authority to manage the American pacification effort. Every program relating to pacification, whether civil or military, now fell under the supervision of Komer and his office. As Westmoreland recalled, it was an "unusual arrangement, a civilian heading a military staff section with a general as his deputy, and a similar pattern of organization was to follow down the chain of command." Thus, the President's "single manager" concept guided reorganization at every level of the U.S. effort in South Vietnam.[22]

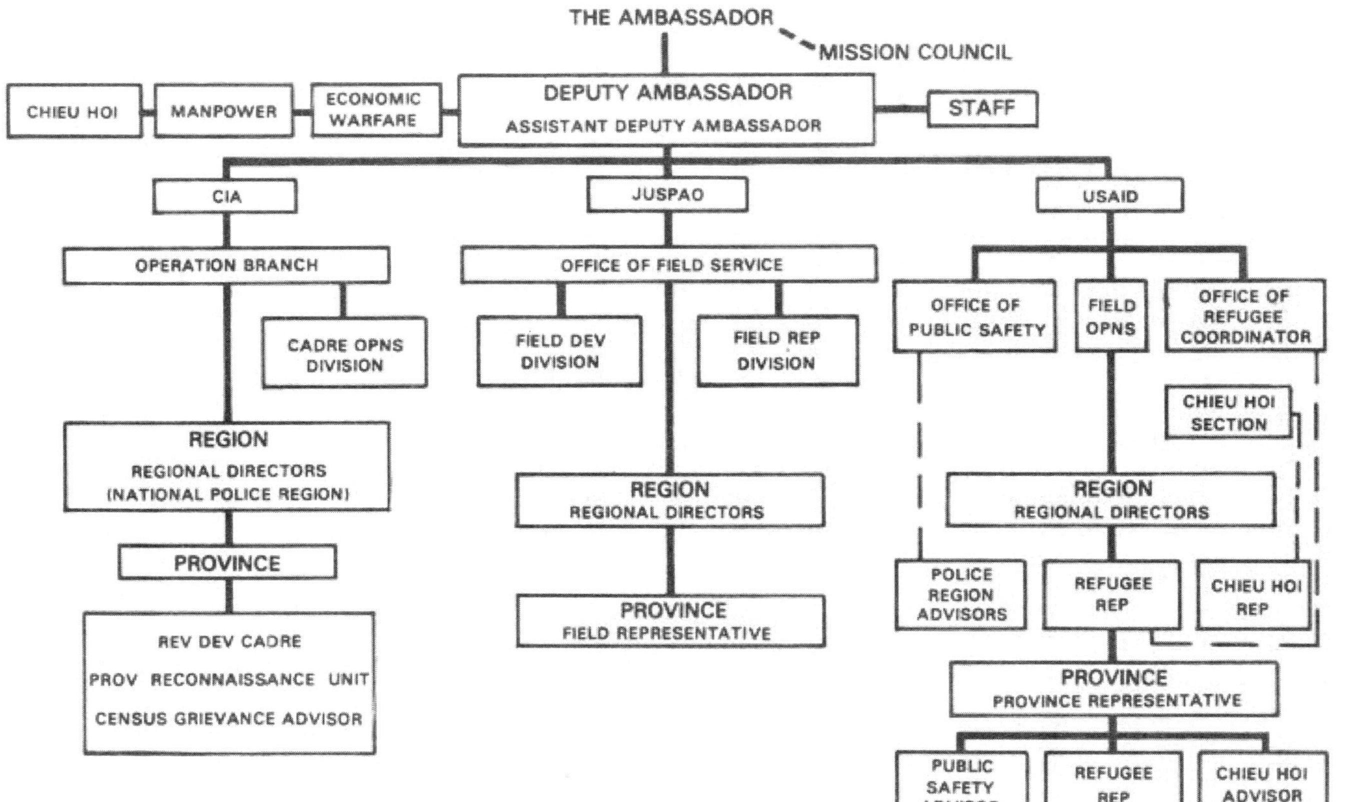

Figure 8-1. Structure of U.S. Mission Showing Position of CORDS, May 1967.

Whereas its predecessor had failed to integrate military operations into an essentially civilian organization, CORDS uniquely incorporated civilians into the military chain of command. The former OCO staff director, a civilian, headed the CORDS office in MACV while a brigadier general served as his deputy. (Komer even received authority for civilians to write performance reports on military personnel.) The main CORDS staff, operating alongside more traditional staff sections like intelligence (J2) and operations (J3), oversaw a wide venue of programs (see Figure 8-2).[23] To make the transition easier, Komer maintained the six field program divisions established under OCO. His reach over pacification programs, however, expanded greatly. "Personnel," Komer recalled, "were drawn from all the military services, and from State, [US]AID, CIA, USIA, and the White House."[24] CORDS assumed responsibility for coordinating rural development programs, conducting village and hamlet administrative training, and overseeing agricultural affairs and public works projects. The integrated, interagency office handled research and development planning, wrote MACV policy directives on pacification, and advised military commanders on civic action programs. Komer even assumed the job of training and equipping South Vietnamese regional and popular forces to provide local security for pacification programs.[25]

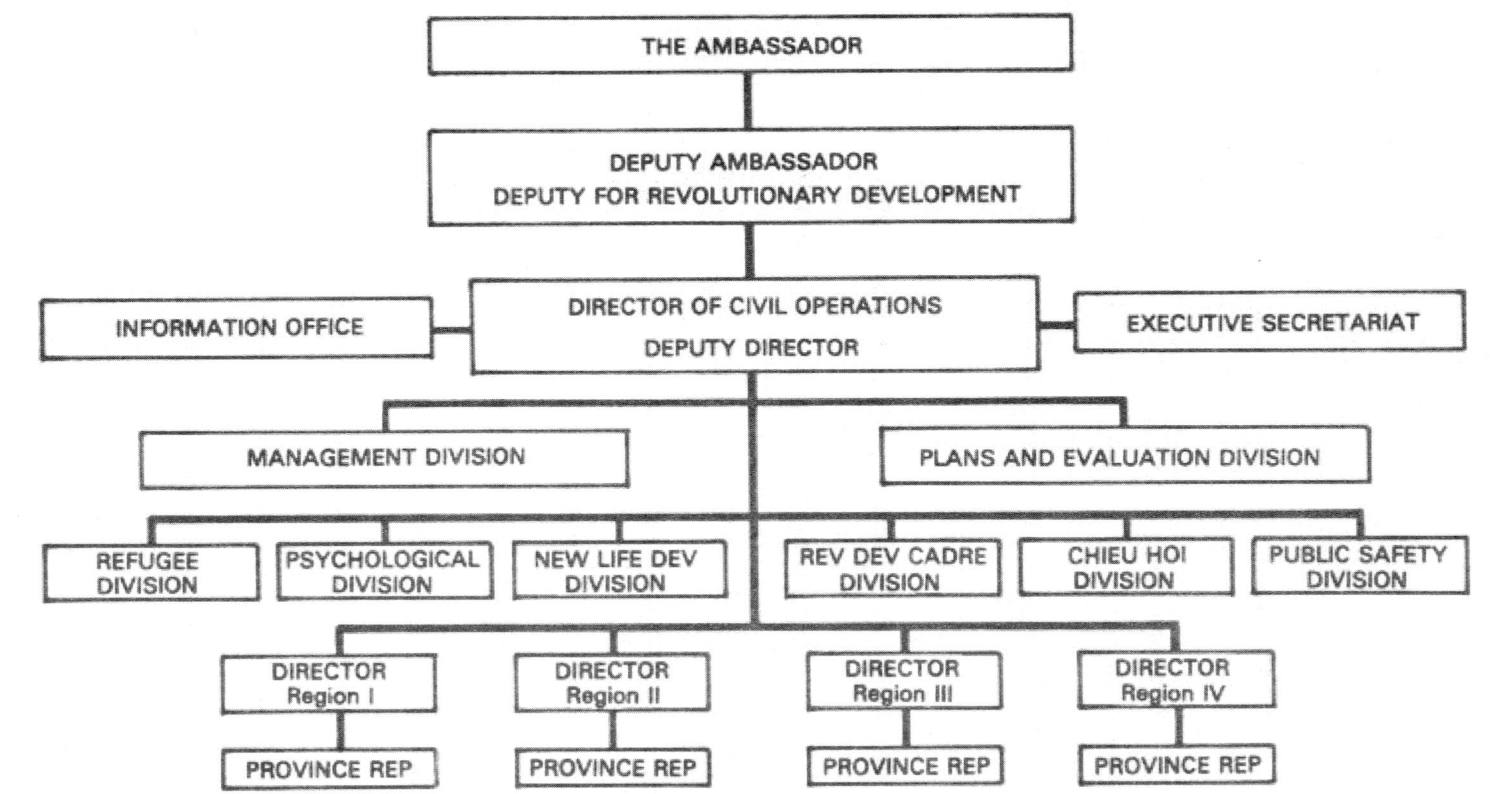

Figure 8-2. Organization of Assistant Chief of Staff for CORDS.

It was here, at the local level, that Komer sought to address the fundamental problems of pacification support through reorganization. The new ambassador assigned each of MACV's corps headquarters a deputy for CORDS, usually a civilian, who outranked the corps commander's chief of staff. Similarly, Komer appointed an advisor to each of South Vietnam's 44 provinces. Illustrating the collaborative approach of CORDS, 25 provincial advisors were military personnel, the other 19 civilians. These province teams reported directly to the corps deputies, while coordinating local military operations with the entire array of pacification programs.[26]

The sheer breadth of pacification requirements, however, strained the capacity of Americans in the field. Reorganization could accomplish only so much. One American colonel, advising a South Vietnamese infantry division, noted the extent of effort required by pacification. Once units had established security, they then had to:

> determine the people's needs, act as a link between the higher governmental agencies and the people, see that the people's needs were met, inform the people, organize hamlet self-government, assist the people in agricultural and economic development, establish intelligence nets, detect and eliminate the Viet Cong infrastructure, and eventually restore the legitimate government in the hamlet.[27]

Establishing a "single manager" for pacification surely made sense. Coordinating and successfully executing the vast number of programs under that manager proved extraordinarily more difficult.

Still, the chief contribution of CORDS was to pull pacification's numerous activities under one centralized command. At its peak, CORDS employed roughly 5,500 officials to support its wide range of programs. Historical precedence in Malaya and external pressure to reform certainly encouraged the reorganization process. So did the support of Westmoreland. The MACV commander gracefully endorsed an arrangement that made few distinctions between civilian and military officials, and he fully backed Komer's ambitions of enlarging the role CORDS played in local population security. As Westmoreland recalled, "Who headed the program at each level depended upon the best man available, not whether he was military or civilian."[28] The MACV commander committed himself to facilitating the implementation of CORDS rather than serving as an obstacle. Thus, if CORDS represented the single most important managerial innovation during the Vietnam War; Westmoreland's support played a decisive role in the organization's inception and subsistence.[29]

THE LIMITS OF A PACIFICATION ORGANIZATION

Innovations in organizational design and management do not lead automatically to innovations in strategic thought or analytic problem solving, nor do they ensure successful execution of programs. CORDS certainly streamlined the process of pacification for MACV but Westmoreland's strategy still required resolving a wide range of military, political, economic, and social problems. Too often in South Vietnam, military operations worked at cross purposes with pacification. Success in one area did not equate to or support advances in the other. Even the metrics for progress to assess population security and pacification efforts proved inadequate, if not inimical, to other efforts under Westmoreland's purview.[30] Moreover, as the MACV 1967 Combined Campaign Plan noted, the "ultimate responsibility for population security in the RD [revolutionary development] plan rested with the RVN [Republic of Vietnam]."[31] The U.S. mission in Saigon may have reformed the American side of pacification, but its South Vietnamese allies, those ultimately responsible for pacification's success, never made comparable revisions. CORDS had solved only half of the problem.

In truth, CORDS never came to grips with the underlying problems of the war inside South Vietnam. If those in the American mission had successfully looked to the past for perspectives on integrating civil-military operations in unconventional warfare, they concomitantly dismissed the failures of past counterinsurgents such as the French in solving intractable problems within South Vietnamese society. Both civil and military leaders assumed that American military power could be wielded successfully to attain any foreign policy objective. In the process, they too often misjudged the difficulties posed by a weak Saigon government. Certainly those involved in pacification recognized the lack of flexibility among their allies, the widespread corruption in both the army and government, and the shortage of initiative and leadership within the South Vietnamese camp. Nonetheless, the CORDS restructuring effort, as significant as it was, never confronted directly these "fundamental constraints on effective administration."[32]

While the American effort in Vietnam ultimately failed to break the enemy's morale, the creation of CORDS demonstrated the Americans' willingness to make organizational changes during a time of war. It equally revealed that American war managers realized success depended on more than just killing the enemy. As Komer recalled, "Pulling together civilian and military efforts . . . led to greater recognition that the war was as much political as military and that adequate interface was essential."[33] Outside political pressure and analogies of past insurgencies surely served as important stimuli for change, but, so too did the readiness of key figures to accept organizational transformations during wartime. This conversion to a unified civil-military staff eventually proved inadequate in furthering American war aims, and perhaps it is here where the CORDS experience offers perspective for those considering organizational change today: True military innovation requires more than simply addressing organizational problems.

ENDNOTES - CHAPTER 8

1. American estimates in Graham A. Cosmas, *MACV: The Joint Command in the Years of Escalation, 1962-1967,* Washington, DC: Center of Military History, 2006, p. 203; and John Prados, *Vietnam: The History of an Unwinnable War, 1945-1975,* Lawrence, KS: University Press of Kansas, 2009, p. 116.

2. Authorization on ground troops in Herbert Y. Schandler, *The Unmaking of a President: Lyndon Johnson and Vietnam,* Princeton, NJ: Princeton University Press, 1977, p. 27. Troop numbers can be found in Robert D. Schulzinger, *A Time for War: The United States and Vietnam, 1941-1975,* New York and Oxford, MA: Oxford University Press, 1997, p. 183.

3. William C. Westmoreland, *A Soldier Reports,* Garden City, NY: Doubleday, 1976, p. 175.

4. As quoted in *Victory in Vietnam: The Official History of the People's Army of Vietnam, 1954-1975,* Merle L. Pribbenow, trans., Lawrence, KS: University Press of Kansas, 2002, p. 154. For an

overview of Hanoi's strategy, see David W. P. Elliott, "Hanoi's Strategy in the Second Indochina War," Jayne S. Werner and Luu Doan Huynh, eds., *The Vietnam War: Vietnamese and American Perspectives,* Armonk, NY: M. E. Sharpe, 1993, pp. 79-83; and William J. Duiker, *Sacred War: Nationalism and Revolution in a Divided Vietnam,* Boston, MA: McGraw Hill, 1995, pp. 179-184.

5. Westmoreland directive, September 17, 1965, in John M. Carland, "Winning the Vietnam War: Westmoreland's Approach in Two Documents," *The Journal of Military History*, Vol. 68, No. 2, April 2004, p. 558.

6. Westmoreland to Brigadier General James L. Collins, Jr., cable MAC 0117, January 7, 1966, Historian's Files, Pacification Overview/Conclusions Folder, Washington, DC: U.S. Army Center of Military History, Fort Leslie J. McNair. (Hereafter cited as CMH).

7. Pacification defined in Larry Cable, *Unholy Grail: The US and the Wars in Vietnam, 1965–8*, London, United Kingdom (UK): Routledge, 1991, p. 132. The essays found in Franklin Mark Osanka, *Modern Guerrilla Warfare,* New York: The Free Press of Glencoe, 1962, are characteristic of contemporary authors—many of them military officers—who sought to use past counterinsurgency operations as a tool for understanding the growing conflict in Southeast Asia.

8. Chester L. Cooper *et al.*, "The American Experience with Pacification in Vietnam, Vol. III: History of Pacification," March 1972, Folder 65, pp. 206-207, U.S. Marine Corps History Division, Vietnam War Documents Collection, The Vietnam Archive, Texas Tech University, Lubbock, TX. Command History, United States Military Assistance Command, 1966, Entry MACJ03, Box 3, RG 472, National Archives and Records Administration, College Park, MD, p. 501.

9. Robert Thompson, *Defeating Communist Insurgency, 1966;* St. Petersburg, FL: Hailer Publishing, 2005, pp. 55. Thompson would later criticize the American effort for its "lack of control" in coordinating pacification programs. See Robert Thompson, *No Exit from Vietnam,* New York: David McKay, 1969, p. 157.

10. Department of the Army, *Field Manual* (FM) *31-16, Counterguerrilla Operations*, Washington, DC: Department of the Army, February 1963, p. 31. On alluring analogies, see Richard E. Neustadt and Ernest R. May, *Thinking in Time: The Uses of History for Decision Makers*, New York: The Free Press, 1988, p. 48.

11. Colonel Wilson quoted in Robert W. Komer, *Organization and Management of the "New Model" Pacification Program – 1966-1969*, Santa Monica, CA: RAND, 1970, p. 17. On lack of coordination among agencies, see John Schlight, ed., *The Second Indochina War: Proceedings of a Symposium Held at Airlie, Virginia, 7-9 November 1984*, Washington, DC: Center of Military History, 1986, p. 127; and George C. Herring, *LBJ and Vietnam: A Different Kind of War*, Austin, TX: University of Texas Press, 1994, p. 69.

12. McNamara quoted in Douglas S. Blaufarb, *The Counterinsurgency Era: U.S. Doctrine and Performance, 1950 to Present*, New York: The Free Press, 1977, pp. 234-235. On the Honolulu Conference, see Cosmas, p. 353.

13. The term "revolutionary development," broadly defined, included the key tasks of population security and nation-building programs. Jeffrey J. Clarke, *Advice and Support: The Final Years, 1965-1973*, Washington, DC: Center of Military History, 1988, p. 172.

14. PROVN report in U.S. Department of State, *Foreign Relations of the United States, 1966*, Vol. IV, Washington, DC: U.S. Government Printing Office, 1985, p. 596. (Hereafter cited as *FRUS*.) See also Andrew J. Birtle, "PROVN, Westmoreland, and the Historians: A Reappraisal," *The Journal of Military History*, Vol. 72, No. 4, October 2008, pp. 1213-1247.

15. Robert M. Montague, "Pacification: The Overall Strategy in South Vietnam," Student Essay, Carlisle, PA: U.S. Army War College, 1966, p. 5, in Robert M. Montague Papers, Box 1, U.S. Army Military History Institute. On Westmoreland's views, see "Command Emphasis on Revolutionary Development/Civic Action Programs," October 22, 1966, in Robert E. Lester, ed., *The War in Vietnam: The Papers of William C. Westmoreland*, Bethesda, MD: University Publications of America, 1993, text-fiche, Reel 7, Folder 10, October 18-29, 1966; "Strategic Guidelines for 1967 in

Vietnam," December 14, 1966, Reel 18, CSA (W.C.W.) Statements Folder, October-December 1966; and U. S. Grant Sharp and William C. Westmoreland, *Report on the War in Vietnam*, Washington, DC: U.S. Government Printing Office, 1969, p. 132.

16. On the single manager concept, see Cosmas, p. 357; for embassy fears, see Herring, pp. 77-78.

17. On OCO evolution, see Richard A. Hunt, *Pacification: The American Struggle for Vietnam's Hearts and Minds*, Boulder, CO: Westview Press, 1995, pp. 82-84; Thomas W. Scoville, *Reorganizing for Pacification Support*, Washington, DC: U.S. Army Center of Military History, 1999, pp. 44-46; and Schlight, p. 131.

18. Westmoreland, *A Soldier Reports*, p. 255. On the inability of a civilian-led program to cope, see Dale Andrade and James H. Willbanks, "CORDS/Phoenix: Counterinsurgency Lessons from Vietnam for the Future," *Military Review*, Vol. 86, No. 2, March-April 2006, p. 12.

19. Cosmas, p. 354; Scoville, p. 54.

20. National Security Action Memorandum No. 362, May 9, 1967, *FRUS*, 1967, Vol. 398.

21. "Structure of US Mission" showing position of CORDS, May 1967, quoted in *The Ultimate Book of Quotations*, compiled and edited by Joseph Demakis, Raleigh, NC: Lulu Enterprises, 2012, p. 50.

22. Westmoreland, *A Soldier Reports*, p. 260. On the "Single manager" concept, see Komer, *Organization and Management of the "New Model" Pacification Program*, p. 55. For background on Komer, see Frank L. Jones, "Blowtorch: Robert Komer and the Making of Vietnam Pacification Policy," *Parameters*, Vol. 35, No. 3, Autumn 2005, pp. 103–118. On CORDS responsibilities, see Hunt, pp. 89-90.

23. "Organization of Assistant Chief of Staff for CORDS," quoted in *The Ultimate Book of Quotations*, compiled and edited by Joseph Demakis, Raleigh, NC: Lulu Enterprises, 2012, p. 50.

24. R. W. Komer, *Bureaucracy Does Its Thing: Institutional Constraints on U.S.-GVN Performance in Vietnam*, Santa Monica, CA: RAND, 1972, p. 115. On CORDS staff, see Scoville, pp. 66-67; and Cosmas, p. 361.

25. CORDS programs in Schlight, p. 133; and Hunt, pp. 90-94.

26. Corps organization in Hunt, p. 94; Civilian and military advisor numbers in Cooper *et al.*, p. 271.

27. John H. Cushman, "Pacification: Concepts Developed in the Field by the RVN 21st Infantry Division," *Army*, Vol. 16, No. 3, March 1966, p. 26. While Cushman's experiences pre-dated CORDS, the requirements of pacification had not changed between 1966 and 1967.

28. Westmoreland, *A Soldier Reports*, p. 260. For numbers of officials in CORDS, see Herring, p. 81.

29. Blaufarb, p. 240; Herring, p. 64.

30. On problems measuring the progress of pacification, see Gregory A. Daddis, *No Sure Victory: Measuring U.S. Army Effectiveness and Progress in the Vietnam War*, Oxford, MA: Oxford University Press, 2011, Chap. 5.

31. Command History, United States Military Assistance Command, 1967, Entry MACJ03, Box 5, RG 472, NARA, p. 317. On CORDS bringing no major conceptual innovations, see Thomas L. Ahern, Jr., *Vietnam Declassified: The CIA and Counterinsurgency*, Lexington, KY: The University Press of Kentucky, 2010, p. 241. Cross purposes within U.S. operations in Marc Jason Gilbert, "The Cost of Losing the 'Other War' in Vietnam," Marc Jason Gilbert, ed., *Why the North Won the Vietnam War*, New York: Palgrave, 2002, p. 179.

32. Cooper *et al.*, p. 273. Similarly, one former intelligence officer complained that "counterinsurgency in Vietnam emphasized military considerations over political ones, enforcement of 'physical security' over more subtle questions of social change and psychological loyalties." David G. Marr, "The Rise and Fall of

'Counterinsurgency': 1961-1964," Senator Maurice Robert "Mike" Gravel, ed., *The Pentagon Papers: The Defense Department History of United States Decisionmaking in Vietnam* [5 Vols.], Boston, MA: Beacon Press, 1971–1972, p. 203. See also CORDS failing to come to grips with fundamental problems in Herring, p. 87.

33. Komer, *Organization and Management*, pp. 246-247.

PART III:

U.S.-ALLIED COMBINED COMMANDS AND ORGANIZATIONS

CHAPTER 9

AN UNQUALIFIED SUCCESS: THE U.S. ARMY AND MILITARY GOVERNMENT IN GERMANY

Kevin W. Farrell

INTRODUCTION

One of the greatest accomplishments in the history of the U.S. Army is its successful occupation of Germany at the end of and immediately following World War II during 1944-49. The reason for this success was early identification of the scope and nature of the mission and selection of the correct personnel well in advance. The relevance of this historic episode for any modern military organization is clear: proper planning and effective leadership are essential to the success of any new command. In addition to its mission of providing military government, the U.S. Army also created a new military organization, the Constabulary Corps, to provide security, stability, and policing functions within the American sector of occupied Germany. The rising tensions of the Cold War and the formal creation of the North Atlantic Treaty Organization (NATO) and the Warsaw Pact prematurely ended the Constabulary Corps as well as the U.S. Army's role in military government in Germany. Ultimately, their striking success and quick demise ensured their obscurity in the historical record. Although largely taken for granted in retrospect, the U.S. Army's unparalleled success stands as a testament to the benefits of good planning and inspired leadership.

One of the greatest success stories in the entire history of the U.S. Army is also one that is rarely remembered or recognized within or outside the U.S. Armed Forces. This oversight is undoubtedly the consequence of that very success, a success that grows evermore remarkable and improbable with the passage of time. Out of the wreckage of a devastated Germany that had spawned the most destructive war in human history causing the deaths of tens of millions of people, the U.S. Army accomplished a feat never before experienced in human history.[1] Although the Allied Powers of Great Britain and France assisted significantly in this process in their respective areas within the "Western zone," it was the U.S. Army that served as the primary agency for not only creating stability, providing security, and disarming the former Nazi regime, but also laying the foundation for one of the strongest and most stable democracies in the world in a region that had never experienced genuinely stable democratic government: the Federal Republic of Germany. An examination of how the U.S. Army accomplished this task is of great relevance today for any command or organization attempting to begin virtually from scratch.

Obviously there were many factors that contributed to this successful outcome while certain circumstances were unique in history (most obviously, the complete devastation and massive occupation of Germany). Above all, however, the keys to the success of the U.S. Army's role in military occupation and military government were thoughtful planning and innovation in execution. Long before there were adequate resources and personnel assigned either to the mission or its future organization, there was frank recognition of the scope of the problem and a general consensus as to how to move forward. This chapter traces how the process went from a vague idea into concrete execution. Additionally, it will briefly explore the genesis of a completely new military organization, the Constabulary Corps.[2]

Almost imperceptibly, the U.S. Army's very success led to swift dismantlement and transformation of this significant achievement in Europe (second in importance only to the role it played in the war). Little over a year after the unconditional surrender of Nazi Germany, the realities of the Cold War necessitated a completely new relationship to the German people and an end to previous occupation policy. On September 6, 1946, in a speech in Stuttgart, Germany, U.S. Secretary of State James Byrnes made it clear that America's approach to Germany had changed from one of occupation to one of protection, proclaiming, "The principal purposes of the military occupation were and are to demilitarize and de-Nazify Germany but not raise artificial barriers to the efforts of the German people to resume their peacetime economic life."[3] He went on to announce a completely new bond between the two countries. Rather than continue in the style of the wartime occupation, Byrnes reassured Germans of the occupation's goal:

> The American people want to return the government of Germany to the German people. The American people want to help the German people to win their way back to an honorable place among the free and peace-loving nations of the world.[4]

This clearly articulated policy anticipated new treaties and finally the creation of a sovereign Federal Republic of Germany: the Brussels Pact of March 17, 1948; the North Atlantic Treaty of April 4, 1949; the approval of the Basic Law and Founding of the Federal Republic of Germany on May 23, 1949; and full membership in NATO on May 5, 1955.[5]

Thus it came to be that shortly after its creation of a military government in Germany, the U.S. Army no longer had a role in occupation *per se* and its post-war military formation dedicated to this task, the Constabulary Corps, had become unnecessary. The U.S. Army continued to have a significant presence in Germany with its force commitment to NATO and continues to have a presence there to this day, although the current role is rather ambiguous, generally consisting of support to NATO and forward staging for ongoing military operations worldwide. Nonetheless, had it not been for the early, spectacular success of the U.S. Army, everything that followed would have been impossible.

The U.S. Army's formal occupation of Germany began before the war ended in the final months of 1944 with a small slice of German territory in U.S. possession in the final months of 1944. The most significant portion under occupation during this period was the historic city of Aachen. Although the U.S. Army had significant experience creating military governments and conducting military occupations in past wars, by the time escalating tensions had led to World War II, these experiences were distant and not well understood or remembered in the U.S. Army. The U.S. Army had played a significant role in the wake of the victory over Mexico in 1847 and 1848 and an even larger role in the occupied Confederate states during the Civil War and Reconstruction. Similarly, during and after the Spanish-American War, soldiers assumed these responsibilities in the Philippines, Puerto Rico, and Cuba. Most recently, and only a generation before World War II, the U.S. Army had occupied part of Germany itself. What all these occupations shared–apart from their much forgotten status–was agreement by the Army and the U.S. Government that such duties were not a legitimate military function.[6] Though relatively limited in scope and duration, the U.S. occupation of Germany following World War I was not recorded as having been a smashing success, and the main official report on it laments, "The American army of occupation lacked both training and organization to guide the destinies of the nearly one million civilians whom the fortunes of war had placed under its temporary sovereignty."[7]

Following the end of World War I, the topic of military government proved to be little more than an abstract concept for the U.S. Army as it shrank dramatically from wartime strength in excess of three million to fewer than a quarter-million soldiers by 1920.[8] The relatively brief military occupation by eight divisions organized into three corps in the Rhineland region of Germany ended on January 24, 1923.[9] With a drastically reduced military in the 1920s and 1930s and very few troops stationed abroad, the issue of occupation and military government quickly fell from mainstream discussion. Only the U.S. Army War College focused any attention on the topic and even then, strictly as matter of law and legal discussion in the classroom rather than an item for which to plan. The few after-action reports from the experience of that war were consistent, in that they all warned against the consequences of lack of preparation and planning.[10] Larger budgetary and manpower issues that further reduced the size and readiness of the U.S. Army throughout the 1930s overshadowed any meaningful reform initiatives.

It is not surprising therefore that during the interwar years, and even at the start of World War II, little serious effort or resources were dedicated to the creation of a military government or occupation organization. This was understandable, considering the U.S. isolationist policies prior to World War II and then the dire situation of the Allies during the early years of the war. Nonetheless, despite having minimal resources and perhaps even lacking enthusiasm for the task, farsighted individuals charged with the mission looked at the scope of the problem as a whole and prepared, as well as they could, for the future.

The main accomplishment during the 1930s was the creation of a field manual that would define the problem and develop a framework for how the army would conduct military government should the situation arise and adequate resources be provided. Long in genesis, two works were indeed published. The first came out in October 1939, published as *Field Manual* (FM) *27-10, The Rules of Land Warfare.* This manual reflected the interwar focus on the legal aspects of military government, although it contained a significant amount focused on civil administration.[11] The second manual proved even more appropriate to the task: FM 27-5, *Military Government,* published in July 1940.[12] In both cases, the international situation added to the urgency of developing an army policy.

By October 1939, Poland had been defeated and invaded completely by Nazi Germany and the Soviet Union, whereas by the summer of 1940, all of Central and Western Europe except Great Britain was firmly under the control of Nazi Germany. With Japan continuing its aggressive imperialism in Asia and Nazi Germany apparently supreme in Europe, it was obvious to American strategists and policymakers that the United States could not remain isolationists much longer. On the other hand, with the growing awareness of how desperately deficient American military preparedness was immediately prior to and after the Japanese attack on Pearl Harbor, HI, on December 7, 1941, the issue of military government was clearly at the low end of the spectrum of priorities. Despite the official state of war against Germany, Italy, and Japan, continued allied setbacks throughout 1942 only reinforced the low priority assigned to planning for military government.

Yet, along with the massive expansion of the Armed Forces that occurred with the first peacetime draft in 1940 and then even more so following the official declarations of war against Japan, Germany, and Italy in December 1941, resources and personnel came far more easily. While resources and personnel improved with each passing month, it is safe to say they never equaled demand until the final months of the war. Key figures in the chain of command, most obviously Army Chief of Staff, General of the Army George C. Marshall, and the U.S. Army Judge Advocate General, Major General Allen W. Gullion, as well as the G-1 and G-3, had the wisdom and foresight to prepare for a future requirement long before it would actually be needed. Their challenge was compounded by the great uncertainties that accompanied the future conduct and indeed the very outcome of the war itself.

By the summer of 1944, when the U.S. Army was firmly established in France, the War Department assigned primary control of the military government mission to the U.S. Army, and the U.S. Army in turn established a two-phased process: the first phase would consist of military government led by the U.S. Army and the second phase would establish a civilian government to replace the military government. However, none of this could have been foreseen clearly in mid-1942. Even so, between September 1942 and June 1944, the G-1 and G-3 sections of the U.S. Army had created the required organizations and training programs in the face of significant bureaucratic opposition. Under the direction of Marshall, the Civil Affairs Division (CAD) was established on March 1, 1943.[13] Major General John H. Hilldring headed the new organization and piloted it through a spirited debate over whether it would report directly to the Secretary of War or to the Joint Chiefs of Staff. While this larger fight was taking place, Hilldring, in trying to build his new command, quickly cited the need for annual training for 1,200 junior officers.[14] Such a figure was openly mocked considering the state of the war at the time, but he and his staff continued to press for the right experts to be drawn from civilian life. As the actual reporting chain of command continued to evolve in the ensuing years and international aspects of the challenge loomed larger as the D-Day landings drew closer, Hilldring and his staff never lost sight of the objective of creating an organization that was staffed by the right people and imbuing in it sufficient flexibility to adapt to changing circumstances.

Even this brief overview provides several key lessons to be gleaned for those charged with the creation of future military organizations and capabilities. First and foremost is a statement of the general issue and ultimate goal. In the case of the occupation of Germany and surrounding territories, the foundation was laid through two seminal documents: FM 27-10, *The Rules of Land Warfare*, and FM 27-5, *Military Government*, which were substantially amended in 1944 with the publication of General Dwight Eisenhower's directive governing the occupation in North West Europe.[15] Even though the details and outcomes were unclear, the U.S. Army leadership had the foresight to identify the general need, and although initially resources were unavailable, save to train a few individuals, the framework would be in place. The second lesson that stands out clearly from the World War II example is the importance of selecting the right personnel for the task at hand, especially in the formative stages. In some cases, the U.S. Army recruited civilian experts and provided them direct commissions at relatively high rank to enable them to act quickly and with sufficient authority. A third key point is that the initial structure built was formed around a core of experts who waited literally for years until their larger command came into existence. By having a central group of experts, the overall organization was infused with flexibility so that, as it expanded, its core competencies were retained. Finally, in order to achieve its ultimate purpose the U.S. Army's occupational effort had to be supported at the highest military and political levels. This demanded that those within the organization were able to express themselves capably to senior army leadership and political masters.

ENDNOTES - CHAPTER 9

1. Earl F. Ziemke, *The US Army in the Occupation of Germany 1944-1946*, Washington, DC: Center of Military History U.S. Army, 1975, remains the best single volume on the topic. It is relatively compact, well written, and comprehensive. Its main shortcoming is the relatively limited period addressed. For a more recent synthesis, see Chap. 14, "Defeat and Occupation," Derek S. Zumbro, *Battle for the Ruhr: The German Army's Final Defeat in the West*, Lawrence, KS: University Press of Kansas, 2006. See also Richard Bessel, *Germany 1945: From War to Peace*, New York: Harper, 2009.

2. Although almost 5 decades have passed since its original publication in 1964, the U.S. Army's extensive planning and initial occupation of Germany and other European countries during World War II is extensively outlined in Harry L. Cole and Albert K. Weinberg, *Civil Affairs: Soldiers Become Governors*, Washington, DC: Center of Military History U.S. Army, 1986. Its successor volume cited in Endnote 1, Ziemke, is of much more practical use for all but the most serious of researchers. Scholarship on the Constabulary Corps is virtually nonexistent, save for a recent release by the Combat Studies Institute at Fort Leavenworth, KS; Kendall D. Gott, *Mobility, Vigilance, and Justice: The US Army Constabulary in Germany 1946-1953*, Fort Leavenworth, KS: Combat Studies Institute Press, 2010.

3. Speech by James F. Byrnes, "Restatement of Policy on Germany," presented in Stuttgart, Germany, September 6, 1946, reproduced on the U.S. Diplomatic Website to Germany, available from *usa.usembassy.de/etexts/ga4-460906.htm*, accessed February 4, 2011.

4. *Ibid.*

5. German Federal Press and Information Bureau, *Facts about Germany*, Frankfurt, Germany: Societäts-Verlag, 2000, pp. 115-120. A major exception to this progression remained the once and future German capital city of Berlin. Berlin remained under Allied military occupation until German reunification in 1990.

6. Ziemke, p. 3.

7. Colonel Irwin L. Hunt, "American Military Government of Occupied Germany, 1918-1920," March 4, 1920, p. 88, reproduced in *ibid.*

8. Richard W. Stewart, ed., *American Military History Volume II: The United States Army in a Global Era, 1917-2004*, Washington, DC: Center of Military History U.S. Army, 2005, pp. 54-56.

9. *Ibid.*, p. 55.

10. Ziemke, p. 3.

11. *Ibid.*, p. 4.

12. *Ibid.*

13. *Ibid.*, pp. 16-18.

14. *Ibid.*

15. Supreme Headquarters Allied Expeditionary Force (SHAEF), "Standard Policy and Procedure for Combined Civil Affairs Operations in North West Europe, London, UK, May 1, 1944.

CHAPTER 10

STANDING UP SHAPE: THE QUEST FOR COLLECTIVE SECURITY IN WESTERN EUROPE

Josiah Grover

INTRODUCTION

On December 19, 1950, President Harry Truman designated General of the Army Dwight Eisenhower as the Supreme Allied Commander, Europe (SACEUR). Eisenhower faced daunting challenges in this new role, not least of which was the development of a joint and multinational command structure to integrate the defense of Western Europe. External pressures from the Korean War and internal political and economic pressures added urgency to the undertaking. Responding to these pressures and the necessity for a speedy activation, Eisenhower enlisted the aid of proven subordinates to stand up a new organization to command the North Atlantic Treaty Organization (NATO) forces in Europe. The Supreme Headquarters, Allied Powers, Europe (SHAPE) evolved as a result of these efforts from December 1950 through SHAPE's activation in April 1951. The challenges for creating a new joint, multinational security command under peacetime conditions were immense and involved complex political-military considerations, questions of jurisdiction, and the use of various prior models to assist in the development of an adequate command structure. Ultimately, the development of SHAPE is a success story, but that outcome was by no means certain in the closing days of 1950. Planners struggled to construct an organizational framework for the new command while simultaneously unifying the military actions of 12 nations. The ultimate solutions resulted from both careful planning and vigorous debate in early-1951.

The morning of December 16, 1950, found Lieutenant Colonel Andrew Goodpaster developing nuclear strategy in the Pentagon, Washington, DC. A decorated veteran of the Italian campaign, Goodpaster had served in the Washington Command Post during World War II, followed by a 3-year stint at Princeton University, Princeton, NJ, for a Ph.D. in international relations. The phone rang that morning with a request from a friend and former supervisor, Colonel Robert Wood. Assuming it a routine matter, Goodpaster agreed to help before asking with what he was going to be helping. Twenty-four hours later, Goodpaster was in Brussels, Belgium, as part of the Advance Planning Group tasked with establishing SHAPE.[1] Far from addressing a single minor issue for an old friend, Goodpaster instead found himself one of the first five officers assigned to a brand new overseas command with no headquarters, grappling with one of the most complex problems American military planners had ever faced in peacetime: how to integrate the military forces of 12 nations under a single commander without the pressure of ongoing combat action to foreclose debate. Their success in doing so rests, in no small part, on the efforts of the Advance Planning Group in developing the command structures for NATO forces and their ability to build a truly joint and multinational headquarters.

THE ORIGINS OF SHAPE – DEFINING THE NEED FOR A UNIFIED COMMAND

That the Atlantic Treaty Nations even undertook such an unprecedented endeavor is worth exploring. On September 26, 1950, the North Atlantic Council approved the "Resolution on the Defence of Western Europe," which called for the "establishment at the earliest possible date of an integrated force under centralized command and control composed of forces made available by Governments for the defense of Western Europe" under the auspices of NATO and commanded by a single Supreme Commander.[2] The initial report by the North Atlantic Council called for a Chief of Staff to stand up a headquarters pending the selection of a Supreme Commander. However, as the international situation deteriorated over the fall of 1950, pressure grew instead to select a Supreme Commander immediately and then worry about the assignment of forces and organization of the command. Continuing uncertainty, driven by the weakness of NATO forces in 1950 and evidence of American military unpreparedness displayed in Korea, contributed to the change. U.S. political and military leaders argued convincingly that naming a commander of some stature could aid in the process of assigning forces to the new centralized command.[3]

Consequently, on October 28, 1950, Truman asked Eisenhower, then president of Columbia University, to visit Washington, DC, to discuss the command of Atlantic Pact forces.[4] The United States had only recently adopted the recommendations of National Security Council (NSC)-68, a document that advocated dramatically increased defense spending and emphasized the military component of Containment.[5] By

December 1950, the Western democracies perceived the Soviet Union and its satellite states as a significant threat, in part because of the ongoing Korean War and the recent intervention of the Chinese Communists in that conflict.[6] Eisenhower, since early November, was fairly convinced that he would soon be requested to command NATO forces and had already begun laying the groundwork for the new command, selecting Lieutenant General Al Gruenther as his Chief of Staff, and requesting other officers.[7] By mid-December it was widely understood in national and international defense circles that Eisenhower would soon be en route to Europe to take command of NATO forces.[8] Truman made it official when he designated Eisenhower SACEUR on December 19, 1950.[9]

IDENTIFYING THE PROBLEMS

As the Advance Planning Group—composed of Gruenther, Wood, Goodpaster, Colonel Dodd Starbird, and Colonel Bob Worden (United States Air Force)—arrived in Europe, they faced a series of challenges; one of the first was finding a place to set up the nascent headquarters. The Hotel Astoria in Paris became SHAPE's first home, offering both rooms to lodge the Advance Planning Group and suitable meeting spaces, while also offering access to communication and transportation networks.[10] The tasks to be dealt with were significant; articulated in the initial planning meeting on December 19, 1950, they included the structure of the major commands; the organization of the headquarters itself; the absence and future integration of non-U.S. officers; the desirability of collecting non-U.S. views regarding the command structure and organization; and the basic problems associated with multiple languages—including how many (and which ones) to use and the viability of using interpreters.[11] These basic problems could be sorted into three categories: the command structure of Allied military forces in Europe; the organizational structure and composition of SHAPE; and the relationship between American and European forces at all levels.

Yet all three categories were inextricably linked and had to be dealt with simultaneously. The Advanced Planning Group embraced a set of basic principles in accomplishing the task, stressing the need to weld separate national air, ground, and naval forces into a unified structure. In order to prevent fissures within the command, different nationalities were to be distributed across the allied command in key positions. Closely related, the boundary between national and NATO missions was deliberately left unresolved in the near term, so as to permit freedom of action in the future. Finally, the planners accepted that resources for accomplishing their tasks were limited, but prioritized the development of command structure as a way to increase available resources.[12] This circular-seeming logic was reinforced by the arrival of Eisenhower in Europe in early January 1951.

Accompanied by his wife, Mamie, Eisenhower commenced a whirlwind tour of the NATO nations, seeking to ascertain the capabilities of each and urging their utmost cooperation in the collective security effort. The unanimous selection to command NATO forces, Eisenhower possessed an enormous amount of prestige, and his presence in Europe bolstered flagging morale. He made explicit use of his stature to secure support for collective security. His efforts to ensure European commitments proved successful, and by the time he returned to the United States at the end of January 1951, he was able to go before Congress and report on significant progress in Europe.[13] While Eisenhower strove to secure concrete commitments in the form of units assigned and funds allotted for collective security, his small staff continued to work on how to integrate those forces into a command structure. They were aided by the work of the Western Union Defense Organization and NATO's three existing Regional Planning Groups, which had done some basic work on organizational concepts for future NATO forces.[14] Upon its activation, SHAPE would essentially replace those four organizations with a single integrated command.

The Allied command structure was, in large part, determined by the geography of Western Europe, stretching from the tip of Scandinavia into the Mediterranean. As such, the planning group determined to split the NATO command structure into three regional commands, North, Central, and South. The North and South commands featured air and naval arenas, flanking the central region, where planners and strategists had determined the decisive clash of arms would likely occur should the Soviets invade Western Europe. In the atmosphere of uncertainty that pervaded Europe in the waning days of 1950, such a possibility was not entirely unrealistic, given what Allied forces knew about Soviet strength in Eastern Europe.[15]

In the North, land forces were split between Norway and Denmark, while Allied air and naval commands were unified in Oslo, Norway. The southern regional command was initially organized in Italy, and featured a single land component, as well as associated air and naval commands. Over time, the southern regional command would evolve new and more complex organizational structures in response to competing national interests and the entry of Greece and Turkey into the Alliance. In the central region, Eisenhower opted to retain control of allied forces directly, with subordinate ground, air, and naval commanders. Eventually, as force structures evolved and grew, a separate central region command would stand up, relieving the Supreme Commander from direct responsibility for that area.[16]

DESIGNING SHAPE: PRIOR MODELS AND INTERNATIONAL CONTINGENTS

Even as Gruenther and the planning group wrestled with forming new command arrangements in Europe, they simultaneously had to develop the structure of the headquarters that would command these new organizations. By mid-January 1951, SHAPE was slowly expanding, no longer restricted to eight or 10 officers as in the first weeks of its existence. Most of the month was consumed by Eisenhower's visits to the member nations and the planning group's attempts to fit the resulting commitments into an organizational framework. While the planners were making progress, in early-February 1951, Goodpaster reported to his friend and mentor, Colonel George "Abe" Lincoln, the head of the Department of Social Sciences at West Point, NY, that:

> we are still in something called the 'Planning Group' rather than 'Headquarters' stage of development, though I must admit we look remarkably like a headquarters of the COSSAC [Chief of Staff to the Supreme Allied Commander] type.[17]

Lincoln, who served as one of General George Marshall's chief planners in World War II and thus knew something of headquarters structure, countered with the observation that SHAPE looked more like the Supreme Headquarters, Allied Expeditionary Force (SHAEF), a comparison others also made.[18] Writing later in 1951, Wood addressed the differences between the new SHAPE and its predecessor:

> there have been some valuable but only rough precedents to SHAPE in the form of wartime combined commands such as SHAEF. . . . The idea of a combined (i.e., more than one nation) command is thus not a wholly new one. What is new is that now 12 nations are involved, and that the undertaking is going forward in peace, rather than being extemporized during war.[19]

The "rough precedents" to SHAPE helped the planning group devise the initial form of the headquarters, which they modeled on the American Army's military staff structure, creating divisions for Personnel, Intelligence, Logistics, Signals, and a G3, which was further divided into Plans, Policy, and Operations on the one hand and Organization and Training on the other.[20] While the basic staff structure was being worked out, the planning group began to transform into a truly multinational organization as the first European officers arrived for service at SHAPE. The addition of international officers complicated the basic functions of the staff, with the addition of new languages and martial traditions; it also, rather paradoxically, relieved some of the pressure on the planning group. By mid-February 1951, Gruenther had imposed a "hiring freeze" for American personnel, especially senior officers, as the accretion of Americans threatened to defeat the intent of a multi-national headquarters.[21] The assignment of European officers brought more perspectives to the planners and rolled back the perception that SHAPE would be an American command with only a token of European representation.

The arrival of the international contingent also explicitly raised the question of national viewpoints. Simply put, SHAPE officers were expected to serve SHAPE first, subordinating their national affiliations to the needs of the headquarters generally and the Supreme Commander specifically. This is not to say that national viewpoints had no representation in the new collective security arrangement—a separate staff liaison group was established specifically to coordinate the national representatives—but individual staff officers, regardless of their national origin or branch of service, were assigned throughout the headquarters according to the needs of the organization and the officer's particular skill set.[22]

While the creation of a separate liaison office for national viewpoints represented one of the most significant early differences between SHAPE and a conventional military headquarters, the distribution of international officers throughout the organization also required a great deal of flexibility. Indeed, as Wood later wrote:

> our methods have undoubtedly seemed very fluid to many of our officers. Particularly, they have been a strain on those officers lacking flexibility and habituated to a full set of regulations, standing procedures and doctrine, the appropriateness of which it was not their task to question.[23]

In fact, an assignment to SHAPE invariably forced officers of all nationalities to grapple with complex problems relating to the integration of multinational and joint forces. The initial SHAPE staff structure had to be flexible enough to cope with problems for which there was no prior precedent, solve the immediate problem, and devise an appropriate mechanism for dealing with similar problems in the future. In the initial phases of establishing the headquarters, that responsibility fell to the newly created office of the Special Assistant to the Chief of Staff held by American Major General Cortlandt Schuyler. In the first 3 months of SHAPE's existence, problems outside the traditional purview of the staff system were resolved at the Special Assistant level, thereby allowing the staff divisions to focus on their particular areas of responsibility.[24]

As the subordinate staff divisions became operational, the international character of the headquarters became more apparent. Of the eight major staff divisions, half were headed by American officers, while the rest were headed by an Italian, two French officers, and one British officer.[25] Deputy Supreme Commander Field Marshall Bernard Montgomery was British, as was the Air Deputy, while the Naval Deputy was a Frenchman. Officers from other nations also filled positions at SHAPE, contributing to the sense that SHAPE was truly an Allied effort.

NEGOTIATING JURISDICTIONS

Even as the multinational character of the headquarters became more evident, the planning group still had to grapple with complex jurisdictional issues, not least of which was the legal role of General Eisenhower. As the SACEUR, Eisenhower had an international mandate to ensure the security of Western Europe, while as a senior American Army officer, he had specific responsibilities and obligations to the American people. This quickly became known as the "dual-hat" problem, requiring that Eisenhower, and his eventual successors, execute both international and American responsibilities.[26] The SACEUR's authority over Alliance military forces raised the specter of an improper American influence in explicitly national matters. From the European perspective, Eisenhower would have to be able to subordinate his American viewpoint to the best interests of the Alliance, while the American public and President could rightly expect him to advance and defend U.S. interests. However, given the SACEUR's role as the commander of military forces from 12 separate nations, acting in response to American interests without regard to those of the other nations, or even **appearing** to act in American interests, would almost certainly damage the SACEUR's personal reputation and diminish the power and prestige of the position. Eisenhower's skill in managing an alliance helped ensure that the issue did not prove particularly contentious during his tenure, as did the fact that American and Alliance interests were closely aligned during that period. Later SACEURs, such as General Matthew Ridgway, would grapple with these two roles, particularly when U.S. and Alliance interests were less convergent.

American congressional and Presidential support assisted Eisenhower in the negotiation of jurisdictional issues. Following his report to Congress in January 1951, the United States designated four Army divisions for deployment to Europe; these divisions would serve as the American military contribution to the collective security of Western Europe.[27] This concrete demonstration of an American commitment to European security helped to mitigate complaints about an American SACEUR.

As the SACEUR and his SHAPE planners overcame the initial obstacles and growing pains of a unique new organization, the commitment of troops to the new regional commands stimulated much debate over the proper sequence of command activation.[28] Ultimately, SHAPE was activated under the authority of NATO when General Eisenhower issued General Order Number One on April 2, 1951. The subordinate regional commands then activated as their headquarters became capable of exercising command and control over the forces allocated to them in the summer of 1951.

ENDNOTES - CHAPTER 10

1. Travel order, FF 13/1, AJG GCMF; letter to Bob Porter, April 26, 1951, FF 12/5 Correspondence-April 1951, AJG GCMF; Andrew J. Goodpaster, "The Development of SHAPE: 1950-1953," *International Organization*, Vol. 9, No. 2, May 1955, p. 257.

2. North Atlantic Council, "Resolution on the Defence of Western Europe" (NATO, September 26, 1950), Document #11 (Final), NATO Declassified Documents, p. 1, available from *www.nato.int/ebookshop/video/declassified/doc_files/C_5-D_11-FINAL.PDF*.

3. Dwight D Eisenhower, *The Eisenhower Diaries*, 1st Ed., Robert H Ferrell, ed., New York: Norton, 1981, pp. 178-179. The argument made was that a caretaker Chief of Staff could only **receive** those forces designated for NATO service, while a Commander could mobilize personal prestige and advocate for forces directly, thereby speeding the transition to an operational command.

4. *Ibid.*, p. 178.

5. George C. Herring, *From Colony to Superpower: U.S. Foreign Relations Since 1776*, Oxford, UK: Oxford University Press, USA, 2008, p. 647. See also the "Surprises" in Memorandum: National Security Planning, George S. Pettee, February 1951, Memorandum for Record, AJG, February 17, 1951, FF 13/3 Official Memorandums-February 1951, AJG Papers, GCMF.

6. Herring, pp. 640-642; Dwight D. Eisenhower, *The Papers of Dwight David Eisenhower*, New York: Columbia University, pp. 1464-1467; Charles M. Spofford, "Toward Atlantic Security," *International Affairs (Royal Institute of International Affairs 1944-)*, Vol. 27, No. 4, October 1951, p. 434.

7. Eisenhower, *The Papers of Dwight David Eisenhower*, Vol. XI, p. 1450. Both Goodpaster and Lieutenant Colonel Pete Carroll were by-name requests; Carroll would serve as Eisenhower's assistant until his death of a heart attack in the White House in 1954, when he was replaced by Goodpaster.

8. C. L. Sulzberger, Special to *The New York Times*, "12 NATIONS CONCUR," *The New York Times* (1923-Current file), December 19, 1950; Drew Middleton, "Why They Picked General Ike," *The New York Times*, (1923-Current file), December 24, 1950.

9. President Truman letter to General Eisenhower, December 19, 1950, FF 12/1-Correspondence-1950, Andrew J. Goodpaster Papers, Lexington, VA: George C. Marshall Foundation. (Hereafter AJG Papers, GCMF.); See AJG Memorandum to Colonel Robert Wood, February 20, 1951, FF 13/3-Official Memorandums-February 1951, AJG Papers, GCMF. The term SACEUR was not actually officially proposed for adoption until February 1951. I use it throughout for the sake of brevity.

10. Welles Hangen, Special To *The New York Times*, "Eisenhower Gets Temporary Office," *The New York Times* (1923-Current file), December 28, 1950.

11. Memorandum for Record, AJG, February 17, 1951, FF 13/3 Official Memorandums-February 1951, AJG Papers, GCMF.

12. AJG, "Principles" note, December 1950, FF 13/1-Official Memorandums-December 1950, AJG Papers, GCMF.

13. James Reston Special to *The New York Times*, "Eisenhower's Magic Wins Over Capitol Hill Suspicion" (1923-Current file), December 1950. See also *NATO, The First Five Years, 1949-1954*, Paris, France: North Atlantic Treaty Organization, 1954.

14. Robert J. Wood, "The First Year of SHAPE," *International Organization*, Vol. 6, No. 2, May 1952, p. 180.

15. Goodpaster, "The Development of SHAPE," p. 255; Spofford, "Toward Atlantic Security," p. 439.

16. *NATO, The First Five Years,* Chap. 4.

17. AJG Letter to G. A. Lincoln, February 12, 1951, FF 12/3-Correspondence-February 1951, AJG Papers, GCMF. COSSAC referred to the Chief of Staff to the Supreme Allied Commander, the precursor to World War II's Supreme Headquarters Allied Expeditionary Force (SHAEF) under Eisenhower. In fact, the selection of Eisenhower as the Supreme Commander of NATO forces was made in part to prevent some of the problems associated with COSSAC, such as troop diversions to other theaters. Operating without full command authority, COSSAC had limited ability to protect its force structure and resources until assigned a commander. See *www.history.army.mil/documents/cossac/cossac.htm,* accessed January 27, 2011.

18. G. A. Lincoln letter to AJG, February 19, 1951, FF 12/3-Correspondence-February 1951, AJG Papers, GCMF.

19. Wood, "The First Year of SHAPE," p. 179.

20. *NATO, The First Five Years*; Goodpaster, "The Development of SHAPE," p. 258.

21. AJG Letter to Edward M. Earle, March 9, 1951, FF 12/4-Correspondence-March 1951, AJG Papers, GCMF; AJG Letter to Major Haakon Lindjord, February 27, 1951, FF 12/3-Correspondence-February 1951, AJG Papers, GCMF.

22. Goodpaster, "The Development of SHAPE," p. 258; Wood, "The First Year of SHAPE," pp. 181-182.

23. Wood, "The First Year of SHAPE," p. 183.

24. AJG Letter to Edward M. Earle, March 9, 1951, FF 12/4-Correspondence-March 1951, AJG Papers, GCMF. The Office of the Special Assistant dealt mostly with American issues, including special weapons and the U.S. Military Assistance Program, which was not technically under Eisenhower's purview but did require some degree of coordination.

25. Wood, "The First Year of SHAPE," p. 190.

26. Major Haakon Lindjord, '*The Problem of Staff Assistance for General Eisenhower's Dual Role*' to AJG, April 1951, FF 12/5-Correspondence-April 1951, AJG Papers, GCMF.

27. C. P. Trussell Special to *The New York Times*, "General Satisfies Congress Critics," *The New York Times* (1923-Current file), February 2, 1951; "Eisenhower Asks Senate Not to Limit Dispatching of U.S. Troops to Europe; U.N. Assembly Votes Red China Guilty," *The New York Times* (1923-Current file), February 2, 1951.

28. See AJG Memorandum to Colonel Robert Wood, February 23, 1951, FF 13/3-Official Memorandums-February 1951, AJG Papers, GCMF; AJG Memorandum to Colonel Robert Wood, March 16, 1951 [SHAPE Activation], FF 13/4-Official Memorandums-March 1951, AJG Papers, GCMF; AJG Memorandum to Colonel Robert Wood, March 17, 1951 [Assignment of Forces], FF 13/4-Official Memorandums-March 1951, AJG Papers, GCMF.

CHAPTER 11

WATCHING THE SKIES: THE FOUNDING OF NORTH AMERICAN AIR DEFENSE COMMAND

Joseph C. Scott

INTRODUCTION

The North American Air (later Aerospace) Defense Command (NORAD) grew out of increasing cooperation between American and Canadian forces in response to the Soviet air threat in the early Cold War. This gradual process was challenged periodically by those opposed to the integration, to American domination of Canadian defense policy, and to strategic air defense in general. Dedicated advocacy by American and Canadian air defense supporters, who took advantage of concerns over world events, eventually won over many of the skeptics. The two nations deployed substantial radar, aircraft, and missile capability against the Soviet threat. To manage this capability, Canadian and American military personnel developed pragmatic, informal solutions to the problems of complex relationships, and this set the stage for more formal bilateral arrangements. As the strategic situation evolved, the organization adapted with it, and NORAD continues an important role to this day.

In August 1957, U.S. Secretary of Defense Charles Wilson and Canadian Defense Minister George Pearkes announced that their respective nations had established "a system of integrated operational control of the air defense forces in the continental United States, Alaska and Canada." The headquarters, to be called "Air Defense Canada/United States" (ADCANUS) would control squadrons of fighter interceptors, hundreds of ground-based anti-aircraft weapons, and a vast, expanding radar network designed to protect the two nations from the threat of the Soviet Union's ever-increasing nuclear bomber capability.[1] Within weeks, the command would abandon its clunky initial title and rename itself the North American Air Defense Command. At the time of NORAD's creation, the defense establishment portrayed the organization as a logical, if crucially important, step in a process dating back to the early years of World War II. As the command and its air defense architecture had developed, however, military officers who supported air defense had faced a number of challenges, not the least of which was skepticism of their ability to defend North America adequately. Throughout NORAD's formative experience and in the intervening years, its advocates overcame such opposition, while also adapting to a constantly evolving strategic arena. NORAD ultimately became one of the world's most enduring bilateral security organizations.

THE EARLY YEARS

In the years immediately following World War II, defense planners paid little attention to the air defense of North America. The Army Air Force established an Air Defense Command in early-1946, but it controlled only two fighter squadrons and a few radars based in and around New York City. With the massive post-war drawdown, meager resources could not support much else in a country that felt it had little to fear because of its geographic isolation and aeronautic and nuclear advantages.[2] Canadian and American air defense planners generated grandiose plans for the defense of the continent, including the establishment of some sort of joint/combined air defense command, but the military leadership of both nations largely ignored these proposals in the austerity of the immediate post-war era.[3]

The gradual escalation in Cold War tensions with the Soviet Union, however, changed this posture. The Berlin Blockade crisis of 1948, and revelations that the Soviets had reverse-engineered the American B-29 bomber into their own long-range bomber led to growing focus on the problem of continental air defense.[4] Initially, the Air Force expanded the number of interceptor squadrons and reactivated old radar systems in Alaska and the Pacific Northwest, but there were concerns that this ad hoc "Lashup" network (named after the engineering term for the temporary connection of equipment for emergency or experimental use) was inadequate against the new Soviet threat. After the Soviets detonated an atomic bomb in 1949, the Air Force secured funding to construct a massive network of 85 radars for Alaska and the Continental United States, along with control centers to coordinate response to a Soviet air attack. To distinguish this system from the "Lashup" network established in response to the crises of 1948-49, the Air Force referred to this new system as the "Permanent System."[5]

The Korean War led to further expansion of American air defense capability, but concerns remained about what to do with America's neighbor to the north. The American defense establishment worried about Canada's limited radar and air defense capability, especially because they figured a Soviet attack would likely come by way of Canada.[6] American and Canadian defensive cooperation had been ongoing since a meeting between American President Franklin D. Roosevelt and Canadian Prime Minister Mackenzie King in August 1940 in Ogdensburg, New York. Concerns over the potential threat that Nazi Germany might eventually pose to the western hemisphere led the two leaders to agree to an organization for studying defensive issues ranging from the German bomber threat to maritime security.[7] This "Permanent Joint Board on Defense" (PJBD) met throughout World War II and continued to assess defense issues in the postwar years.

EXPANSION AND CONTROVERSY

Thinking that Canadian radar capability was critical to American national security because of the early warning it could provide, the United States paid two-thirds of the cost for the construction of new air defense radars in southern Canada and near American "Lend-Lease" Air Force bases in Newfoundland starting in 1951. The initial plan for this new "Pinetree" radar system was for the Royal Canadian Air Force (RCAF) to man almost all of the stations, but the Canadian military lacked the manpower to operate all 33 radars. The Americans agreed to man half of the radars, and attempts to keep the decision quiet failed, resulting in public controversy in Canada over the growing American military presence on Canadian soil. Headlines lamented Canada's fate as "another-Belgium" caught between the United States and the Soviet Union. RCAF personnel reassured many doubters that Canadian sovereignty was not in jeopardy, and over the next few years, the two air forces directly coordinated procedures for intercepting and destroying enemy aircraft, including those instances where either nation's interceptors had to cross the international border.[8] Informal cooperation between the U.S. Air Force (USAF) and RCAF became a routine part of air defense planning and operations.

Expanding both nations' interceptor and radar capability was a good start, but defense planners remained concerned about the ability of the growing air defense system to defend North America. The USAF Chief of Staff publicly acknowledged that the technology of the day could hope to destroy only 25 percent of incoming Soviet aircraft and "undoubtedly" it would be "less than that."[9] In the age of atomic and thermonuclear weapons, such a figure was wholly unacceptable, and the Air Force sought answers from the scientific community.

Two groups of scientists came up with proposals to improve the continent's air defenses, focusing on the development of computerized management of air defense operations: improvements to fighter and surface-to-air missile defenses, and the extension of the radar network to the far northern reaches of Canada. Significant portions of the American armed forces, including some within the Air Force, opposed this expansive vision for continental air defense because they believed it would hamper growth and improvements to the nation's offensive nuclear capability and its conventional forces. Supporters of strategic air defense won the day: the Dwight Eisenhower administration's shift to the "New Look" national security policy in 1953 led to a massive buildup in the Strategic Air Command's (SAC) nuclear arsenal, but it also gave the green light for expansion of the nation's air defense posture.[10] The United States and Canada agreed to build two lines of radar stations, one across 55 degrees north latitude (funded and manned by

Canada) and one along the northern shores of arctic Canada (funded and manned by the United States).[11] Construction of the "Mid-Canada Line" and the "Distant Early Warning (DEW) Line" began in 1955.[12]

THE BIRTH OF NORAD

By the late-1950s, air defense radars, aircraft, and surface-to-air missiles positioned throughout the continental United States, Alaska, Canada, and even Greenland formed a complex web of command and control relationships. While the USAF maintained the lion's share of the responsibility, the Army's ground-based air defense mission and the Navy's potential role in guarding the eastern and western approaches to the United States meant they also played a part in continental defense, and there were ongoing squabbles over issues of command and control between the three services. Early ad hoc solutions paved the way for the more formal joint Continental Air Defense Command (CONAD), which was granted operational control over the air defense systems and units of all involved services. The Army Air Defense Command (ARADCOM), for example, which was greatly expanding as it deployed dozens of Nike missile batteries throughout the United States and overseas in places such as Thule, Greenland, now fell under the overall control of the Air Force general who commanded CONAD. ARADCOM, like its Navy counterpart, established headquarters co-located with CONAD's base in Colorado Springs, CO.[13]

The command and control relationship was even more complicated in Canada. The two nations' services coordinated so that U.S.-manned radar stations and American fighter squadrons operating in Canada (either because they had crossed the border or because they were based at U.S. facilities in Newfoundland) were under the operational control of the RCAF. The only stipulation was that the Canadians would exercise this control through the senior U.S. commander present whenever possible.[14]

The ongoing radar expansion and the continued close relationship between U.S. and Canadian air forces increasingly intertwined the defense postures of the two nations. Military and civilian air defense advocates on both sides of the border began seeking a more formal structure to coordinate continental defense, but these efforts faced resistance from both Canadian and American military and political authorities. The senior leadership in Canada remained hesitant about subordinating their military independence to the Americans, and when the RCAF's chief of staff publicly referred to an official joint American-Canadian air defense command as "inevitable" in 1955, the Canadian defense minister quickly repudiated the claim. Admiral Arthur N. Radford, the Chairman of the U.S. Joint Chiefs of Staff echoed this criticism. The heads of the American services were concerned that an official agreement with the Canadians would limit American freedom of action, especially because Canadian discussions about a combined command had borrowed heavily from the precedent set by the North Atlantic Treaty Organization (NATO), of which Canada and the United States were both members. The Joint Chiefs had no interest in subordinating another military command to the whims of NATO.[15]

The air defense leadership within the RCAF and USAF got around these hurdles by proposing, through the American defense establishment, that American and Canadian air defense efforts be "operationally integrated" without forming an actual combined command. The Canadian military agreed to the plan, and a combined study group developed the concept of "Commander-in-Chief Air Defense Canada-United States." The commander in chief (CINC) ADCANUS and his staff would develop procedures for the operational control of the air defense forces commanded by the two respective nations, but they would not be called a command. The study group proposed that the American commander of the Continental Air Defense command become CINCADCANUS, and that the commander of the Canadian air defense forces serve as the deputy.[16] The Joint Chiefs, Canadian chiefs, and the American secretary of defense all approved the proposal. In the summer of 1957, after a brief delay due to political changeover during Canadian parliamentary elections, the Canadian minister of defense also endorsed the concept. At roughly the same time, the DEW Line in northern Canada became fully operational.[17]

Secretary of Defense Wilson and Defense Minister Pearkes announced the new institution on August 1, 1957, but within weeks, General Earl Partridge, the first commander of this not-quite-a-command, requested that its name be changed, and that the parties involved simplify the awkward nature of the institution by acknowledging that it was, in fact, a command.[18] Both the Canadian and American military chiefs agreed, and the organization's name was changed to the North American Air Defense Command. NORAD was formally activated on September 12, 1957.[19]

The new command, based in Colorado Springs, CO, tied together the Continental Air Defense Command and its subordinate headquarters (the respective Air Force, Army, and Navy air defense components) with the RCAF Air Defense Command. As under the original model, the CINC of NORAD was an American and the deputy a Canadian.[20] Within weeks of NORAD's activation, however, opposition to this arrangement emerged within the Canadian parliament. The two militaries had founded this unified command outside diplomatic channels, effectively giving command of significant Canadian forces to an American (and, of course, theoretically giving command of American forces to a Canadian, if the senior officer was absent). The fact that there was no formal agreement between the two heads of state on the issue concerned many Canadians.

In a move that shocked the American Joint Chiefs, the Canadian Prime Minister persuaded the opposition in Parliament of NORAD's merits by arguing that NORAD was essentially an extension of NATO. While the original August 1957 announcement had included a portion on bilateral air defense cooperation extending "the mutual security objectives of the North Atlantic Treaty Organization to the air defense of the Canada-US Region," the Joint Chiefs were concerned that tying NORAD explicitly to NATO would open the door for European interference with North American air defense.[21] Behind the scenes, Canadian military leaders assured their American counterparts that they had no intentions of tying the new command to NATO, and the secret "Terms of Reference" for NORAD contained no references to NATO.[22] While the two militaries negotiated with this understanding, their political masters and diplomats developed a formal agreement containing allusions to NATO, but in vague enough terms that the respective militaries could be confident that NORAD would effectively remain separate from NATO.[23] The vague references were enough to convince virtually all of Canada's lawmakers to agree to the new formal agreement.[24] NORAD now traces its official founding to this second, formal agreement, dated May 12, 1958.[25]

NORAD EVOLVES

NORAD had survived interservice and diplomatic concerns during its creation, but new challenges arose almost immediately. Skeptics complained about the redundant staff structures and headquarters of the joint American Continental Air Defense Command and NORAD at Colorado Springs. General Partridge responded by proposing to deactivate CONAD because it seemed unnecessary after the establishment of NORAD, especially since the Canadians had no equivalent organization. The Joint Chiefs of Staff refused Partridge's proposal, but he and his staff simply combined the NORAD and CONAD headquarters and eliminated all separate CONAD staff structures.[26]

The changing strategic environment was a more serious challenge to NORAD. By the late-1950s, effective intelligence gathering, including flights by U-2 reconnaissance planes over the Soviet Union, revealed that the Soviets had far fewer long-range bombers than had been feared.[27] Increasingly, nuclear planners and theorists on both sides of the Iron Curtain were emphasizing the role of intercontinental ballistic missiles (ICBMs) as nuclear delivery systems. American intelligence experts believed the Soviets would have a "significant ICBM capability" by 1960. NORAD's current radars and anti-aircraft weapons would be ineffective against this new threat. Even as the American government was negotiating the formal NORAD agreement with its Canadian neighbors, the United States was moving towards augmenting the air defense radar systems in northern Canada and Greenland with the Ballistic Missile Early Warning System. The commander of NORAD worked to ensure these new radars would fall under NORAD's purview.[28]

Intelligence revelations and the shift to ICBMs meant that the halcyon days of massive continental air defense buildup were over. Starting in 1959, less than 2 years after NORAD became operational, Congress began chipping away at air defense budgets.[29] By the late-1960s, the continental air defense force structure was a shadow of its former self.[30] The Navy and then the Army air defense commands deactivated as their missions vanished.[31] NORAD retained a key role, however, through its continued use of the radar networks in the far north of the continent. For the rest of the Cold War, its primary mission was to warn the SAC of an impending Soviet attack so that SAC could launch its counterstrike.[32]

NORAD was the result of a gradual process of coordination and cooperation fostered from below by air defense advocates in both the United States and Canada. Bit by bit, these service members developed pragmatic solutions to complicated issues involving joint and combined cooperation, diplomacy, and incredibly complicated technological systems. They overcame opposition from within their own services, from other military branches, and from their own governments to gather allies and advocates, and ultimately established a vast network of radar systems and weapons to defend North America from atomic attack, and the organization to control all this. As the strategic and threat situation evolved, so too did North American air defense efforts.

ENDNOTES - CHAPTER 11

1. "U.S. and Canada to Combine Air Defense in One Command," *The Washington Post and Times Herald*, August 2, 1957.

2. NORAD Command History Division, *Historical Review of North American Aerospace Defense, 1946-1970*, Historical Reference Paper No. 15, Colorado Springs, CO: NORAD, 197, p. 1. (Hereafter cited as *Historical Review*.)

3. Joseph T. Jockel, *No Boundaries Upstairs: Canada, the United States, and the Origins of North American Air Defense, 1945-1958*, Vancouver, British Columbia, Canada: University of British Columbia Press, 1987, pp. 19-20.

4. *Historical Review*, p. 2; David Cox, *Guarding North America: Aerospace Defense During the Cold War 1957-1972*, Peterson Air Force Base, CO: NORAD Office of History, p. 4.

5. *Historical Review*, pp. 2-4; Jockel, *No Boundaries Upstairs*, pp. 32-34.

6. Jockel, *No Boundaries Upstairs*, pp. 43-44.

7. For a detailed discussion of the Ogdensburg agreement, see Stanley W. Dziuban, *The Relationship between the United States and Canada, The United States Army in World War II*, Washington DC: Center of Military History, 1990, pp. 22-30. The text of the declaration is on p. 24.

8. Jockel, *No Boundaries Upstairs*, pp. 24, 44-54. On the number of stations in the "Pinetree" radar system, see *Historical Review*, p. 4. "Pinetree" was chosen because it started with the same letter as "Permanent."

9. General Hoyt S. Vandenberg, quoted in Cox, p. 6.

10. "A Report to the National Security Council by the Executive Secretary on Basic National Security Policy," NSC 162/2, October 30, 1953, pp. 5-6, Digital National Security Archive, accessed February 2, 2011.

11. Jockel, *No Boundaries Upstairs*, pp. 79-83.

12. Cox, p. 9.

13. *Ibid.*, pp. 28-29; *Historical Review*, p. 6.

14. Jockel, *No Boundaries Upstairs*, pp. 93-95; and Cox, pp. 28-29.

15. Joseph T. Jockel, *Canada in NORAD, 1957-2007: A History*, Montreal, Quebec, Canada: McGill-Queen's University Press, 2007, p. 22.

16. *Ibid.*, pp. 22-24.

17. *Ibid.*, pp. 24-25; and *Historical Review*, p. 13.

18. Jockel, *Canada in NORAD*, p. 25. Newspaper articles covering the announcement had even referred to the organization as a "command." See, for example, *The Washington Post* and *Times Herald*, August 2, 1957. Jockel implies that Partridge was also concerned about the last four letters in "CINCADCANUS."

19. *Historical Review*, p. 35.

20. *Ibid.*, pp. 25-26, 36.

21. Jockel, *Canada in NORAD*, pp. 30-35. Quote is from *The Washington Post* and *Times Herald*, August 2, 1957, p. A1.

22. Jockel, *Canada in NORAD*, p. 34. See also, "Revised Terms of Reference for the Commander in Chief, North American Air Defense Command," May 7, 1958, Digital National Security Archive, accessed December 22, 2010.

23. "Texts of Statement and U.S.-Canadian Notes on Air Defense," *The New York Times*, May 20, 1958.

24. "NORAD Voted in Canada: Commons Favors Air Defense Plan with U.S., 200-8," *The New York Times*, June 20, 1958.

25. North American Aerospace Defense Command, *Guarding What You Value Most*, Washington, DC: Government Printing Office, 2008, p. 8.

26. Jockel, *Canada in NORAD*, pp. 26-27.

27. James Meikle Eglin, *Air Defense in the Nuclear Age: The Postwar Development of American and Soviet Strategic Defense Systems*, New York: Garland Publishing, Inc., 1988, p. 159.

28. "Report to the President by the Security Resources Panel of the ODM [Office of Defense Mobilization] Science Advisory Committee, Ballistic Missile Early Warning System (BMEWS) for ICBM," March 28, 1958, Digital National Security Archive, accessed December 22, 2010. "ICBM capability" date is from Enclosure "C."

29. Eglin, pp. 181-184, 198.

30. Michael Getler, "Big Cut Set in U.S. Air Defense Force," *The Washington Post* and *Times Herald*, October 7, 1973.

31. North American Aerospace Defense Command, *Guarding What You Value Most*, pp. 11, 14.

32. *Ibid.*, p. 1.

PART IV:

DEPARTMENT OF DEFENSE AND CABINET-LEVEL ORGANIZATIONS

CHAPTER 12

A HISTORY OF NO HISTORY: THE INTELLECTUAL ORIGINS OF THE AMERICAN AIR FORCE, 1917 TO 1948

Gian P. Gentile

INTRODUCTION

The history of the origins of the American Air Force is, interestingly, a history of a rejection of history. That is to say, as the early thinkers and organizers of the American air arm began to emerge in the waning years of World War I and into the 1920s, they eschewed the history of war as irrelevant for them and their budding service. American airmen, like many of their European counterparts, believed that the airplane was a radically transformative piece of technology and that its use in military operations had revolutionized war itself. Because the military history of past wars was confined to fighting on the ground or at sea—fought without airplanes—airmen considered that history to be irrelevant. How could history be relevant, they thought, if there were no airplanes involved? Discarding history, however, caused serious problems, and over the next 20 years—as American air forces were established—created conceptual holes in their thinking and led them to misunderstand the nature of their new weapon, its potential effects, and its overall role in warfare. These problems often manifested themselves in constructed metaphors and analogies that tended to confuse and misguide rather than to explain and inform. Military history, if they had paid attention to it, might have prevented them from making these mistakes.

American airmen started to consider aerial warfare soon after the Wright Brothers made their first flight, and, like warfare more generally, it encompasses a variety of complex and interrelated ideas. Even crafting a logical and generally agreeable definition of "aerial warfare" is difficult and reveals some of the problems in approaching the subject ahistorically. First, aerial warfare can be understood as a category of warfare. It is the art, or activity, of conducting war from the air. More specifically, it could be defined as the application of destructive force from things that fly through the sky. Indeed, in war in general, many things fly through the sky to kill, or destroy, or observe other things on the ground.

In 1225, at the famous battle of Agincourt in the Hundred Years War, King Henry V's English longbowmen loosed flurries of arrows that flew through the sky and struck their French enemies. Surely, the French considered these objects falling from the sky to pierce their armor dangerous, yet, clearly, this was not aerial warfare. Thus, one thing that might seem to make aerial warfare distinct is that the things that fly through the sky, be they bombs, bullets, or missiles, are launched from a platform that is also in the sky. Over time, those platforms have included kites, hot air balloons, and airplanes; today the possibility exists of using satellites in outer space. But even this definition is not air tight. Indeed, missiles launched from the ground that travel across oceans to attack targets—referred to as intercontinental ballistic missiles—are also considered part of aerial warfare. Common definitions of aerial warfare also include the reconnaissance and observation of enemy forces and nations. Generally speaking, "aerial warfare" is the conduct of war from the sky against other objects in the sky and objects on land or at sea as well, either to destroy or to provide information.[1] As the preceding discussion demonstrates, the history of aerial warfare is closely linked to the history of ground and sea warfare; ignoring this connection defines the subject too narrowly.

The first flight of an airplane by two American brothers, Orville and Wilbur Wright, on December 17, 1903 in Kitty Hawk, NC, was the start. The Wright Brother's invention had immediate, and obvious, practical military utility. In fact, when *The New York Times* broke the story, it reported that the Wright Brothers were in a hurry to sell their "airship" for "scouting and signal work, and possibly torpedo warfare" to the government.[2] Word of the Wright Brother's invention quickly spread, and within 5 years, the two brothers were traveling to Europe giving demonstrations of winged flight. All of the major European armies, as well as the Japanese army, understood the military significance of manned aerial flight through the use of the airplane and began developmental programs in their armies to build air forces.

But it was World War I, which began in August 1914 over a fragile balance of power between the major European states and would ultimately involve countries the world over that realized the potential of the airplane as a primary agent in the conduct of war. World War I ushered in many of the ideas, methods, applications, and technological developments of aerial warfare. Historian Stephen Budiansky rightly notes, "the effect of the First World War on military aviation was greater than the effect of military aviation on the war." What is meant here is that airpower—a new term that came into widespread use by airmen during World War I and which seemed to encapsulate a more inclusive conception of the many different forms of warfare in the air—did not win the war singlehandedly, nor did it win any battles on the ground, but it did prove the critical military importance and utility of airplanes and other machines that flew through the air. Aerial warfare as it evolved in the 20th century can trace its roots directly back to World War I.[3]

The development of fighter aircraft and the training of pilots to protect reconnaissance aircraft produced, by the middle part of the war, an idea among airpower advocates that was to have huge consequences during the remainder of the war and beyond: air superiority—or as Italian airpower theorist Guilio Douhet called it shortly after the war, "Command of the Air." If one side's airplanes could completely control the skies either by shooting down the opponent's aircraft or keeping them on the ground and thus preventing them from flying at all, it could then do whatever it wished from the air. If it controlled the skies, it could conduct reconnaissance and observation, direct artillery, and strike at targets such as railroads and even cities which lay deeper in the rear of the field armies. Airpower advocates came to believe that command of the air (i.e., operate without interference from enemy aircraft) was required to have the necessary freedom of action to pursue their strategic aims.[4]

As they looked to build their own air arm, American airmen, such as American flyer and combat veteran from World War I Brigadier General William (Billy) Mitchell, bought into Douhet's ideas. For them, command of the air did many things for an air force. For one, to achieve air superiority was to give *raison d'être* for a call for independence from the Army they supported. They argued that in order to establish command of the air, airmen had to be left unfettered by what they saw as the stodgy ways of the ground-bound land armies. There was no place for the history of past wars fought on the ground for airmen like Mitchell and Douhet. Since the history of war prior to the development of the airplane was naturally about ground armies and navies in combat, American airmen like Mitchell thought it constrained the potential of the airplane and airpower because it would make them subordinate to ground commanders. This was anathema to them.[5] So instead of placing the airplane in the context of greater military history, they started from a conceptual scratch that drew on analogies and metaphors to explain how airpower should be used and the war winning effects it could create.

It was out of this intellectual ferment that airmen began to look to the strategic bombing of cities and other sources of enemy power as a mission fitting for independent air forces.[6] In the minds of many American airmen who flew and fought during World War I, the use of the airplane in that war established the immense promise of airpower—namely, that airpower alone could be sufficient to achieve victory in war. Of these thinkers and practitioners, Italian theorist of airpower Douhet became the most prominent. In the 1920s, Douhet argued that the airplane had revolutionized warfare and that, if used and applied correctly, it could eliminate ground warfare in the future. In his 1921 book, *Command of the Air*, Douhet, for example, reflected on trench warfare in World War I and then imagined the next war in which massive fleets of bombers flew over armies (making them irrelevant in his mind) and going directly after vital population centers in enemy cities. Douhet stated that the past history of warfare provided no guide for war in the future revolutionized by the airplane and airpower. Yet, by jettisoning the past through history, Douhet came to base his theory of airpower on some very dubious assumptions.

The first and most important of these faulty assumptions was that a fleet of airplanes dropping bombs on enemy cities and causing terror would necessarily cause that enemy population to force its government to capitulate.[7] His views were not uncommon among air zealots. Mitchell, for example, believed that the American military must transform itself around the war-winning potential of the airplane that could conduct strategic bombing raids against enemy cities and industry. Both of these men thought World War I and the use of airplanes in it pointed to a future dominated by aerial warfare.[8]

During the years between the end of World War I and the outbreak of World War II, American airmen developed a doctrine referred to as "high altitude, daylight precision bombing." American airmen believed that if the right targets were chosen and bombed effectively, then an enemy nation's "industrial web" could be broken and the nation possibly even forced to surrender. The term "industrial web" was a metaphor that represented a complex, modern nation-state and its economy and infrastructure. Embedded within this metaphor, however, was the theoretical underpinning to the doctrine of "precision bombing." The metaphor likened the modern state's economy and infrastructure (the industrial web) to the literal web a spider spins and concluded, therefore, that if the industrial web were hit at certain key points, the entire web would collapse.

American airmen constructed a theory, which they argued would work in practice, that fleets of American bombers flying over enemy cities could drop bombs on critical targets and destroy them, and from this tactical process of destruction, the enemy state would collapse and ultimately surrender.[9] A better understanding of history might have suggested otherwise, but then again, American airmen, as they began to build their air arm in the 1930s, were not really interested in what they could learn from history. In order to put this theory into practice, however, American airmen understood that they needed to have an independent air force, unfettered by ground officers wanting to use air assets in support of operations on land. To be sure, there was logic to this argument, but American airmen pushed their theory and doctrine into almost a faith in order to argue for independence. The faith's central tenet, then, was the wholehearted acceptance that airplanes could destroy a so-called "industrial web" through bombing, and the result would be the collapse of the enemy state.[10]

If this airpower religion were, in fact, true, then it made perfect sense to establish an independent air arm. However, the establishment of an independent air arm would have implications for military and national strategy in war, since a large independent air arm would take scarce resources away from the other military arms. In the run up to World War II, American airmen were unsuccessful in gaining full independence from the Army, but they had come a long way toward achieving their goal by the start of World War II because, in reality, the air arm within the Army was relatively independent, with its own organizational structure and chain of command. With an air campaign plan to attack Nazi Germany called Air War Plans Division-1 (AWPD-1), airmen garnered from American President Franklin Roosevelt a substantial commitment in American resources to build an independent fleet of bombers that would put into practice the theory and doctrine developed in the 1930s. But would American "precision bombing" work in World War II?

As the United States entered the war, it would be severely tested. In addition to the theoretical underpinning of the industrial-web concept, American airmen also made a series of important tactical assumptions, namely that their bombers could fly during the day and protect themselves from enemy fighters with their large amounts of protective machine guns on board. This assumption was deadly. In August 1943, after about a year of building up its bomber fleet to attack German industry, the American Air Forces took such serious losses from German fighters and anti-aircraft fire from below that they had to reassess their operations fundamentally. From that month on, American bombers usually had fighter protection as they flew.

As the war went on, American airmen were always hopeful that their bombing campaigns against Germany and Japan would produce victory—namely the unconditional surrender of both states—through airpower alone. With regard to Germany, this did not happen, and it ultimately took the combinations of aerial bombardment by American and British air forces with ground operations in France and to the East, and then into the heart of Germany to bring about German surrender. Certainly, airpower played a critical role in the defeat of Germany, but it did not do it by itself as the originators of the "industrial web" theory had hoped. Japan, and its surrender, presented a less clear cut case as to what ended the war in the Pacific.[11]

Indeed, as World War II came to an end in the Pacific in August 1945, nearly 4 months after it ended in Europe, American airmen began to argue that airpower delivered by the Army Air Forces (AAF) largely alone had defeated Japan and brought about its unconditional surrender. A major report sponsored by the AAF and produced at the end of the War, the *United States Strategic Bombing Survey*, argued that British and American strategic bombing did not end the war on their own in Europe, though it had contributed greatly to that end. But with regard to Japan, the survey's authors argued in a breathtaking conclusion regarding the cause for Japan's unconditional surrender that even if the United States had not dropped two atomic bombs on Hiroshima and Nagasaki, and even if a land invasion had never been planned or contemplated, Japan still would have surrendered due to the effects of the American strategic bombing campaign against Japanese cities.[12]

So, in the pages of the *Strategic Bombing Survey*, American airmen and their civilian proponents sought to create a fresh history of the end of World War II in Japan as the result solely of American airpower. Other events that played a crucial war in Japan's surrender—like the Navy's successful campaign to wreck Japanese commerce from the sea, the threat of a major American land invasion, and the Soviet entrance into the war—were all buried under a newly-created history that sought to place the cause of American victory over Japan on the backs of American bombers. But the actual history of why the war ended in Japan was much more complex than the simplistic notion that American bombers pummeling Japanese cities had won the war. In the next 2 years, from 1945 to 1947, American airmen used these sorts of conclusions from the *Strategic Bombing Survey* to help them gain congressional approval to establish an independent AAF in 1947.[13]

American airmen began their crusade for an independent Air Force after World War I with a rejection of history outright. Then, at the end of World War II, they constructed a fresh history from scratch, but that scratch-made history ignored the greater causes and complexities of ending the war in the Pacific. This flawed history, on which the AAF was founded, would haunt the United States in the years ahead as it foolishly guided an American air effort in the Vietnam War that hoped to break Vietnamese communist will through aerial destruction. As was the case in the past, in the absence of solid historical reasoning and military analysis, American military leaders in Vietnam fell back on the use of metaphors to try to explain what they were doing. Air Force Chief of Staff and former World War II American bomber leader General Curtis LeMay said that, in order to defeat the communists, the United States should bomb North Vietnam "back into the stone age."[14] LeMay's quip betrays the theoretical underpinnings of American airpower that went back to the notions of "industrial web" in World War II in that by breaking North Vietnam's "industrial web," it would be placed back into the "stone age" and hence would surrender to American demands. But the seduction of metaphor in place of sound historical reasoning prevented airmen like LeMay from realizing that as far as industrial power went, North Vietnam was already in the "stone age" at that time.

Still, the use of metaphors instead of sound historical thinking would continue. The next metaphor that clouded clear military thinking based on flawed notions of the history of airpower would be the belief in 2003 at the beginning of the second American War in Iraq that American airplanes dropping bombs and missiles from the sky would "shock and awe" Iraq into submission. Again, faulty historical thinking that produced understanding through metaphor prevented a clear-eyed view of what airpower could really accomplish in war, and more importantly its limits.

ENDNOTES - CHAPTER 12

1. Stephen Budiansky, *Air Power: The Men, Machines, and Ideas that Revolutionized War, from Kitty Hawk to Gulf War II*, London, UK: Viking, 2004, pp. 7-10.

2. Peter L. Jakab, *Visions of a Flying Machine: The Wright Brothers and the Process of Invention*, Washington, DC: Smithsonian Press, 1990, pp. 48-49.

3. Budiansky, pp. 125-128.

4. Guilio Douhet, *The Command of the Air, Translated by Dino Ferrari, 1942*, Washington DC: Office of Air Force History, 1983, pp. 3-5, 25-27.

5. David MacIsacc, "Voices from the Central Blue: The Air Power Theorists," Peter Paret, ed., *Makers of Modern Strategy*, Princeton, NJ: Princeton University Press, 1986, pp. 629-632.

6. William "Billy" Mitchell, *Winged Defense: The Development and Possibilities of Modern Air Power – Economic and Military*, New York: Kennikat Press, 1971, p. 215; and Alfred F. Hurley, *Billy Mitchell*, Midland, TX: Midland Books, p. 103.

7. Mitchell, *Winged Defense*, pp. 164-166.

8. Douhet, pp. 3-5, 15-24; Thomas M. Coffey, *Hap: The Story of the U.S. Air Force and the Men Who Built It*, New York: Viking Press, 1982, pp. 277-278.

9. Peter R. Faber, "Interwar US Army Aviation and the Air Corps Tactical School: Incubators of American Airpower," Phillip S. Meilinger, ed., *The Paths of Heaven: The Evolution of Airpower Theory*, Maxwell Air Force Base, AL: Air University Press, 1997, pp. 219-225.

10. Robert Futrell, *Ideas, Concepts, Doctrine: Basic Thinking in the United States Air Force, 1907-1960*, Maxwell Air Force Base, AL: Air University Press, 1989, pp. 108-113.

11. Conrad C. Crane, *Bombs, Cities, and Civilians: American Airpower Strategy in World War II*, Lawrence, KS: University of Kansas Press, 1993, pp. 117-119, 138-142; Tami Davis Biddle, *Rhetoric and Reality in Air Warfare: The Evolution of British and American Ideas about Strategic Bombing, 1914-1945*, Princeton, NJ: Princeton University Press, 2002, pp. 270-288.

12. *United States Strategic Bombing Survey*, Chairman's Office, Summary Report (Pacific Report), Washington, DC: U.S. Government Printing Office, 1946, p. 26.

13. Gian P. Gentile, *How Effective is Strategic Bombing: Lessons Learned from World War II to Kosovo*, New York: New York University Press, 2001, pp. 131-132; and Barton Bernstein, ed., *The Atomic Bomb: The Critical Issues*, Boston, MA: Little, Brown, 1976.

14. Mark Clodfelter, *The Limits of Air Power: The American Bombing of North Vietnam*, New York: Free Press, 1989, pp. x; and Curtis E. LeMay with MacKinlay Kantor, *Mission with LeMay*, Garden City, NY: Doubleday, 1965, p. 565.

CHAPTER 13

TRIAL AND ERROR: THE CREATION OF THE NATIONAL SECURITY AGENCY

Kevin A. Scott

INTRODUCTION

The conflicts of the early-20th century showed the American government the value of collecting communications intelligence (COMINT) on the enemies, and sometimes the friends, of the United States. Interservice rivalries and the lack of collection coordination initially hampered this effort. However, the global nature of World War II forced the services to tentatively deconflict their COMINT collection responsibilities. The U.S. Government also provided its military with incredible amounts of resources to accomplish this mission. This situation changed significantly at the end of World War II. With rapid demobilization came reduced resources to continue the COMINT mission. At the same time, more and more government agencies requested access to intelligence about the Soviet Union. President Harry Truman signed the National Security Act of 1947 in part to coordinate COMINT collection. The act created the United States Communications Intelligence Board (USCIB) as a mechanism for synchronization between America's COMINT collectors and its consumers, but the system did not work smoothly because the USCIB did not have the authority to compel the cooperation of its constituents. In 1949 the Armed Forces Security Agency (AFSA) was established in an attempt to solve some of these problems by merging the COMINT sections with the armed forces, but this military solution ignored the needs of civilian agencies. This fact, combined with significant intelligence failures during the Korean War, prompted Truman to create the National Security Agency (NSA) in 1952, with the express mission of controlling the COMINT collection for the U.S. Government.

In the Presidential Directive that created the NSA in 1952, COMINT was defined as "all procedures and methods used in the interception of communications other than foreign press and propaganda broadcasts and the obtaining of information from such communications by other than the intended recipients."[1] Although the United States already conducted rudimentary COMINT in each of its conflicts, this field did not really come into its own until the widespread use of radios in the early-20th century made the interception of enemy transmissions relatively easy. But this ability to read enemy messages did not translate directly into an effective collection program because it was so resource intensive to keep up with the increasing sophistication of the methods used to protect these messages. American COMINT changed during the 20th century to meet the challenges posed by each new national conflict, and the trend during this period was toward centralization. The U.S. Government brought most of its communications assets together with the creation of the NSA in 1952. Truman ordered the creation of the NSA to reinforce the primacy of civilian control over the COMINT apparatus and to streamline the use of intelligence resources.

The technology that allowed for the rapid transmission of messages over great distances developed in the late-19th and early-20th centuries. The Great War (World War I) taught American military leaders the importance of securing their own long-range communications and the advantages that could be gained by intercepting the internal messages of their adversaries.[2] Following World War I, the U.S. Government relied upon the Army and the Navy to conduct the COMINT mission. While the services had the equipment and manpower to accomplish this task, it also proved to be another venue for intraservice competition. As an NSA history of the period notes, "each [service] worked . . . on those military and naval targets of direct interest to themselves."[3] Army and naval COMINT assets were not always employed against corresponding targets, which prevented them from creating a holistic intelligence picture. The services also used their monopoly on COMINT to intercept the diplomatic cables of foreign governments, resulting in duplicated efforts as both services attempted to impress the President and the Secretary of State by being the first to update them on such transmissions.[4] Thus, interservice competition and an unclear mission for the military COMINT collection and analysis programs in the 1920s and 1930s resulted in both inefficient collection of intelligence and an incomplete intelligence picture.

This parochialism began to change during World War II when the demands of a two-front war forced the Armed Services to begin working together to manage the collection of COMINT. In June 1942, representatives of the Army, Navy, and Federal Bureau of Investigation (FBI) met to plan the COMINT focus of each of their agencies and, by extension, the U.S. Government. Under the pressure of fighting in both the Atlantic and the Pacific Theaters, the Navy agreed to let the Army collect all diplomatic traffic until the end of the war.[5] This marked the first time that either service agreed to cede any primacy in the politically critical realm of diplomatic intelligence. The attendees also recommended that other governmental agencies should be prohibited from collecting COMINT to prevent the expansion of this key field outside the control of military and law enforcement authorities.[6] They wanted to keep COMINT collection and management capabilities within the agencies best prepared to conduct this mission and to exploit the information provided. Though this meeting was successful and the services were willing to compromise in order to complete the wartime mission, the COMINT environment was still inherently competitive. The services did not want any more competition in the COMINT field, and the President agreed for the duration of the war.

The Armed Forces' ability initially to integrate and then cooperate in the collection of intelligence contributed to the successful exploitation of enemy communications throughout World War II. Among other achievements, the Army successfully deciphered Japanese diplomatic traffic, while the Navy decoded many Japanese and German naval messages.[7] The services effectively established a connection between COMINT collection and battlefield success. As a result, they expanded their COMINT collection and analysis operations. Recognizing the utility of working together, and not wanting to provide room for bureaucratic competitors, the services created the Army-Navy Communications Intelligence Board (ANCIB) in 1944 to further facilitate collaboration and to arbitrate any disagreements.[8] By the end of the war, the uniformed services had enough men, equipment, and money to maintain COMINT coverage across most of the world.

The end of World War II, however, affected American intelligence collection in conflicting ways, essentially decreasing resources across the board, but increasing the number of agencies demanding a role in collecting COMINT. First, as a result of Truman's order for rapid demobilization, an 80 percent reduction in the size of the U.S. military created shortages in both manpower and experience that rendered worldwide coverage impossible.[9] Second, other governmental agencies, unwilling to adhere to the wartime precedent of deferring COMINT collection to military ownership, claimed a role in the process. These agencies, which had become heavily reliant on COMINT products during and after the war, wanted to shape collection priorities in addition to accessing products already collected by the now understaffed military collectors. By 1946, both the State Department and the FBI had joined the once exclusively military ANCIB, now known as the USCIB. The military did, however, maintain control over the national COMINT focus because it still owned the collection assets. This arrangement led to growing tension between the civilian federal agencies and their uniformed counterparts, as both desired detailed intelligence about their own specific national missions.

These de facto cooperative policies were transformed into a more organized collection strategy in 1947, when the President signed the National Security Act, as part of a broader program intended to provide coherence to the government's defensive posture in the post-war world and to reassert civilian control over the American military. The act established the National Security Council (NSC), whose mission was to advise the president about national security matters and to promote cooperation among governmental departments.[10] In its role as primary coordinating agency, the NSC began issuing Intelligence Directives

(NSCIDs) to identify the specific missions of each intelligence organization and to prevent any overlap or redundancy. Issued in March 1950, Directive Number 9, "Communications Intelligence," added the newly-formed Central Intelligence Agency (CIA) to the USCIB membership and gave the USCIB the mission to affect "the authoritative coordination of Communications Intelligence activities of the Government."[11] Recognizing an increased role for other nonmilitary, federal agencies in national defense, the act widened participation in the USCIB and added additional oversight from the NSC. This structure added more civilian oversight to the military collection program and attempted to increase the efficiency of intelligence sharing within the government, thereby meeting the growing demands for information and reducing tensions between intelligence consumers.

Since many considered the Soviet Union (then the Union of Soviet Socialist Republics [USSR]) the primary threat to the United States and its foreign policy, the USCIB focused the American COMINT collection assets against the USSR. The Army and Navy had early success intercepting and decoding Soviet governmental communications, but Soviet defensive measures including landlines and cipher systems made COMINT collection increasingly troublesome, if not impossible. Additionally, each armed service maintained separate collection efforts against their Soviet sister service in addition to the other requested targets coming from the USCIB, which reignited the interservice competition that had characterized COMINT collection prior to World War II. In 1947, the creation of the U.S. Air Force (USAF) only exacerbated this problem. The fledgling Air Force Security Service demanded that it be allowed to gather COMINT about the Soviet strategic bomber force, but without the manpower, budget, or expertise to do so, the chief of USAF intelligence admitted that his organization's COMINT ability was severely lacking.[12] The military's inability to produce valuable COMINT due to a shortage of trained specialists and collection equipment led to complaints throughout the government and from the USCIB itself.[13] To address these issues, Secretary of Defense Louis A. Johnson issued Joint Chiefs of Staff (JCS) Directive 2010 on May 20, 1949, establishing the AFSA. Its mission was to "conduct all communications intelligence and communications security activities within the Department of Defense [DoD]."[14] He hoped that unification would force the services' cryptologic units to work together more efficiently to provide a better product.

As it looked for more effective ways to manage the collection of COMINT, DoD studied several foreign examples. The first system they analyzed was the contemporary, integrated Soviet cryptographic program. Advocates of centralization argued that the unified Soviet communications program allowed them to more effectively utilize all of their resources and expertise by avoiding the duplication faced by American officials.[15] Others looked to the most recent experience of worldwide conflict, World War II, to understand the most useful way to build a global intelligence collection network. They asserted that decentralization hampered German COMSEC (communications security) while consolidation facilitated British communications interception and decoding efforts.[16] For them, the events surrounding the Allied cracking of the Enigma code and the subsequent inability of German intelligence to recognize this situation supported their push for American COMINT centralization.

These examples supported the drive within the DoD to consolidate its intelligence structure in the form of AFSA. They did little, though, to identify the proper role for civilian agencies in the COMINT collection and analysis process.

Though it represented progress in synchronization of the military collection effort, the creation of AFSA also complicated the bureaucratic controversy characteristic of the ongoing rivalry for primacy in directing COMINT. The civilian agencies each thought that AFSA's new mission conflicted with the mission of the USCIB.[17] They did not object to the goals of efficiency and economy espoused by the Secretary of Defense, but they did fear losing the ability to provide input concerning the conduct of COMINT collection. The Director of the AFSA reported to the Joint Chiefs of Staff. Because of this relationship, the civilian agencies suspected that the JCS alone was establishing the COMINT priorities for the new agency and the military collection assets that belonged to it; a task that they claimed rightfully belonged to the USCIB. For the civilian agencies, this arrangement contradicted the intentions of the 1947 National Security Act and NSCID 9.[18] Johnson had also gotten presidential approval for the merger without the consent of the USCIB. This end-run infuriated the nonmilitary agency representatives and earned him their mistrust.

The outbreak of the Korean War in June 1950 tested the ability of the new American intelligence community to respond effectively to international crisis. Structural problems inherent in the design of the national COMINT apparatus, particularly the overlap and resulting tension between AFSA and USCIB, became apparent with the outbreak of the Korean War on June 25, 1950. The North Korean invasion took the American intelligence community by complete surprise. The first problem was that AFSA was under-resourced and, as a result, the majority of U.S. COMINT intercept and processing personnel were assigned to Soviet communications, which were considered to have collection priority.[19] In fact, staffing levels were so low that many targets were given a lower priority or, in many cases, ignored. This lack of focus on credible, but non-critical target areas like North Korea meant that, at the time of the invasion, "there were no traffic analysts working on North Korean communications, no Korean linguists . . . and an almost total absence of knowledge of North Korean terminology."[20] For an American military used to communications superiority, such a failure was shocking. This lack of resources was acknowledged in a July 1950 Joint Chiefs memorandum that directed that AFSA receive 1,961 additional COMINT personnel and an $8.13 million dollar increase in Fiscal Year 1951.[21] Such increases were welcome, but having lost many trained and experienced personnel to post-war staff reductions, this infusion of people and money would take time to achieve meaningful results.

The second problem highlighted by the Korean War was the ambiguous nature of AFSA's relationship with the tactical COMINT units of the military services. The AFSA charter "excluded from AFSA's control those COMINT facilities and activities that served in direct support of the field commanders for the purpose of providing tactical intelligence."[22] This exclusion limited AFSA's ability to intercept strategic communications because they were forced to rely on the same collection assets as the military's tactical commanders. This issue became more complex once the North Koreans instituted more stringent communications security on their higher level traffic. This transition eliminated the little strategic intelligence that AFSA had been intercepting. Tactical intercepts became the only source of COMINT, and both AFSA and the field commanders wanted priority in analyzing and exploiting it. This limited the effectiveness of AFSA in the eyes of other governmental agencies that did not have ready access to tactical COMINT of the military services and reignited the old collection rivalries.

The intelligence failures of the Korean War left AFSA and military dominance within the COMINT system vulnerable to critics who advocated a new COMINT strategy. Walter B. Smith, Director of Central Intelligence, proposed a review of Communications Intelligence activities to the NSC. In a memorandum dated December 10, 1951, he stated that:

> it is believed that existing means of control over, and coordination of, the collection and process of Communications Intelligence have proved ineffective. . . . Because of the unique value of Communications Intelligence, this matter directly affects the national security.[23]

In response, Truman ordered the Secretaries of Defense and State and the Director of Central Intelligence to form a special committee to survey the national COMINT structure.[24] They formed the Brownell Committee, consisting of representatives from the State Department and the CIA. Significantly, the military was excluded in the review process.[25] This circumstance could not help but telegraph the conclusion that both the President and civilian authorities were dissatisfied with military efforts at planning and synchronization and therefore sought to reorganize these systems to civilian specifications. Ultimately, the mistrust created by this unilateral change under Johnson ensured that the civilian agencies would not seek input from the military branches as they designed its successor.

First, the civilian agency-dominated Brownell Committee focused on the government's institutional intelligence relationships. They identified COMINT resources as "being national in nature."[26] This was a critical source of intelligence, which was very important to more than just the military. Even though the civilian agencies had a vote on the U.S. Communication Intelligence Board, the military controlled the actual collection by having AFSA report directly to the JCS. To make COMINT more responsive to agencies outside the military, Brownell's committee advocated that the Secretary of Defense, rather than the JCS, directly control AFSA. For the first time, a civilian would be directly responsible for the organization that collected the COMINT. They also put forth the idea that the USCIB be reorganized and given "greater responsibility for 'policy and coordination' in COMINT matters."[27] These changes would allow for more oversight and information sharing as part of a truly national COMINT effort.

Second, the committee determined that AFSA's authority should be broadened in order to streamline its efforts and increase productivity. The committee concluded that the problem of insufficient resources resulted from the fact that AFSA had "insufficient authority or control over the COMINT activities of the three services."[28] Without the ability to force compliance from the service's COMINT activities, the AFSA director had to waste time and effort negotiating with the service representatives for collection assets needed to fulfill the mission. Therefore, the committee recommended that "AFSA should have operational and technical control over all the COMINT collection and production resources of the military services."[29] This would not only save AFSA officials time but also avoid the duplication of intelligence effort across the armed services. The second problem the committee noted was a transient workforce. A lot of time and money went into training COMINT collectors and analysts. However, the report noted a high rate of turnover due, in large part, to "the lack of professional and managerial opportunities for the civilian workforce."[30] Civilians felt excluded from decisionmaking roles in these primarily military organizations. To remedy this, Brownell's committee suggested that AFSA open up more leadership paths to civilian employees.

The President agreed with most of the Brownell Committee's recommendations and issued a Directive on October 24, 1952, entitled "Communications Intelligence Activities." As recommended, Truman ordered the creation of the newly-named NSA. To reassert civilian control of communications intelligence, he identified the Secretary of Defense as "executive agent of the Government, for the purpose of COMINT information."[31] The USCIB was also given a clearer mandate and all civilian agencies received two votes on the board as opposed to each military service's one. This directive also clarified that the military's COMINT services worked for the NSA director without the qualifications that had hampered AFSA leadership. Civilian concerns were addressed by mandating that the military director "have a civilian deputy whose primary responsibility shall be to ensure the mobilization and effective employment of the best available human and scientific resources."[32] In this manner the Presidential Directive reasserted civilian control of COMINT and streamlined the use of limited governmental resources.

The first director of the NSA, Army Lieutenant General Ralph Canine, also worked to win the trust of his civilian workforce. To maintain the secret nature of this new organization, DoD planned to establish the NSA headquarters at Fort Knox, KY. This disturbed a larger portion of the cryptologic workforce who lived in and around Washington, DC. Canine petitioned the Secretary of Defense to keep the NSA close to the capital for the principal purpose of mollifying his civilians.[33] This action, combined with new opportunities for advancement, demonstrated the importance that NSA leaders placed on their combined military and civilian workforce.

These policy changes at both the national and agency levels reflected the growing importance of civilians to the national defense following the end of World War II. Nonmilitary federal agencies demanded more say in the decisions regarding the targeting and exploitation of COMINT. Civilian specialists also wanted more recognition of their skills and the ability to move into leadership positions. These trends, combined with a military establishment looking to maximize the use of their finite resources, led to the centralization of American COMINT in the period between World War II and the Korean War. This evolution can be seen in the NSA's original 1952 structure. While imperfect, the NSA initial organization was the successful product of hard-earned experience and bureaucratic maneuvering.

ENDNOTES - CHAPTER 13

1. Edward C. Keefer, Douglas Keane, and Michael Warner, eds., *The Intelligence Community, 1950-1955*, Foreign Relations of the United States, 1950-1955, Washington, DC: Government Printing Office, 2007, p. 353.

2. John Keegan, *Intelligence in War*, New York: Vintage Books, 2004, p. 193.

3. Thomas L. Burns, "The Quest for Cryptologic Centralization and the Establishment of NSA 1940-1952," *Series V: The Early Postwar Period*, Washington, DC: U.S. National Security Agency, Center for Cryptologic History, VI, 2005, p. 5, available from *www.nsa.gov/about/_files/cryptologic_heritage/publications/misc/quest_for_centralization.pdf*.

4. *Ibid.*, p. 6.

5. *Ibid.*, p. 9.

6. *Ibid.*

7. *Ibid.*, p. 15.

8. "The Origins of NSA," Washington, DC: U.S. National Security Agency, Center for Cryptologic History, p. 2, available from *www.nsa.gov/public_info/_files/cryptologic_histories/origins_of_nsa.pdf*.

9. Matthew M. Aid, *The Secret Sentry*, New York: Bloomsbury Press, 2009, p. 9.

10. National Security Act of 1947, Public Law 80-235, *U.S. Statutes at Large* 61, 1947, p. 495.

11. Glenn W. LaFantasie, David S. Patterson, and C. Thomas Thorne, Jr., eds., *Emergence of the Intelligence Establishment, 1945-1950*, Foreign Relations of the United States, 1945-1950, Washington, DC: Government Printing Office, 1996, p. 1123.

12. Aid, p. 20.

13. *The Intelligence Community*, pp. 22-23.

14. Burns, p. 49.

15. *Ibid.*, p. 51.

16. *Ibid.*, p. 53.

17. Burns, p. 49.

18. *Ibid.*, p. 57.

19. Aid, pp. 21-22.

20. Burns, p. 70.

21. *The Intelligence Community*, pp. 26-27.

22. Burns, p. 71.

23. *The Intelligence Community*, p. 229.

24. "The Origins of NSA," p. 4.

25. Burns, p. 83.

26. *Ibid.*, p. 87.

27. *Ibid.*, p. 88.

28. *Ibid.*, p. 86.

29. *Ibid.*, p. 87.

30. *Ibid.*, p. 88.

31. *The Intelligence Community*, p. 350.

32. *Ibid.*, p. 354.

33. James Bamfort, *The Puzzle Palace*, Boston, MA: Houghton Mifflin Company, 1982, pp. 58-59.

CHAPTER 14

THE DEPARTMENT OF HOMELAND SECURITY

Matthew J. Flynn

INTRODUCTION

The following analysis of the Department of Homeland Security (DHS) targets issues and challenges confronting the department as it was "stood up" in an atmosphere of crisis in the wake of the September 11, 2001 (9/11) attacks. DHS is portrayed as a department struggling to complete its mission without adversely affecting civil liberties at home, all while requiring military assets to protect the homeland from terrorist attacks and to mitigate the effects of natural disasters. On the positive side, DHS is seen as setting a new path in terms of defining military actions and capabilities in an area germane to, and intimately linked to, the civilian sphere. This chapter is presented with the caveat that the publications discussing DHS are remarkably limited, and rectifying this deficiency will be essential to telling the full creation story of DHS.

Formed in the aftermath of the tragedy of 9/11, DHS is a key security organization, which, as of this writing, is still developing a command structure. The key question is how (and how effectively) DHS managed unprecedented challenges in terms of "standing up" a command. Because of a lack of available DHS source information, however, it is difficult to pinpoint key players and measure results and achievements. While its relative newness within the defense system means a lack of historical perspective and concrete conclusions about its formation, it also makes it a compelling case study precisely because it is so new. Indeed, in a changing world featuring terrorism and related threats, DHS may be a useful model to study as new commands learn how best to balance military and civilian concerns in terms of planning operations, understanding the effect of technology on the security apparatus, and gauging whether the scope of such challenges are beyond a single agency's ability to address. This last point is unsettling, for if a department cannot complete its mission, how then does one cope with the changing world and the resultant dangers that affect national security? The DHS currently wrestles with this very problem, and if it does so imperfectly, it is certainly not for lack of attention to the issue. Rather, the primary challenge for DHS—and, indeed, for any command that must fully integrate civilian and military components—is that it must seek security in a primarily civilian setting without curbing foundational American civil liberties.

ORIGINS

DHS faced structural command challenges from its inception. There are two main reasons for this. First, those responsible for shaping the department had to calculate whether a domestic response to terrorism should include both a civilian and military component and, if so, in what mix. Second, decisionmakers feared the department was too large, with too much bureaucracy that could render it ineffectual. These concerns grew out of the crisis-driven context of the creation of DHS, coming as it did in the wake of the 9/11 terrorist attacks. These two fundamental problems were not resolved early on, and DHS faced reorganization after only a year in existence.

The sense of all those involved in the formation of DHS was that the new Cabinet-level department was essential to face a new threat, that of terror. What this organization most represented to its creators was a response to a new age, one dominated by terrorist threats, and therefore the DHS was upheld as the "cutting edge" answer to this new era, at least on the home front. In this view, there were (and are) no analogies to past organizations or commands that could be made, since it represented a fundamental departure from the past.

Just how "new" this all was, however, is questionable. The terror attacks of 9/11 may have been new in how they were delivered, but external threats to American security obviously have always existed. So, for DHS, the key was to ask whether it was in danger of overkill when it came to creating a new bureaucracy to respond to the threat. This was, especially in an era that was imbued with a conservative mantra of not "growing" government, a deeply political question related to long-standing debates within the American system about the appropriate level of government presence at home. Thus, the fear of a garrison (militarized) state, honed during the Cold War, permeates this discussion, and it does so for good reason. The question is, essentially, this: In response to a terrorist attack on the home front, whether by an internal or external enemy, what is the required response? Does it require a military response, civilian response, or any change at all at home? In sum, are law enforcement agencies obsolete in this new age of terrorism, and if they are, is the DHS equipped to fill the void?

Those involved in standing up DHS, and presumably sustaining it now, declared that a change was necessary. The assumption behind the entire edifice of

DHS is that 9/11 was possible to a large degree because of the failure of government agencies to work together and share information. Infamously, and underscoring this problem, many of the terrorists perpetrating the attacks had trained at American flight schools, and the Federal Bureau of Investigation (FBI) was aware of this **before** the attacks occurred.[1] The implication is that more effective communication within the government may have thwarted the attackers. So, better coordinating the government's response became the initial mandate of DHS, one involving almost 180,000 government employees and billions of dollars. The urgency of the situation lent credence to the belief that a better home front defense could be accomplished with the vast government apparatus already in place if only that apparatus were better managed. In the rush to form DHS, this basic assumption went unquestioned.

Initially, the department consisted of five directorates: Border and Transportation Security; Emergency Preparedness and Response; Science and Technology; Information Analysis and Infrastructure Protection; and Management. Under Tom Ridge, the first head of DHS, it became a viable agency in this form on January 24, 2003.

FROM RIDGE TO CHERTOFF

Circumstance soon forced the DHS to redefine itself as something more than an organization designed to stop terrorist attacks on U.S. soil. Hurricane Katrina's devastation and destruction of New Orleans, LA, prompted this realignment of purpose in August 2005. "Homeland security" would now include a broader mission of emergency relief and disaster response. Michael Chertoff, the second head of the DHS, recast the Department's mission statement with this dual purpose made clear:

> pursue a unified, risk-managed strategy of preventing or reducing America's vulnerability to terrorism and natural disasters . . . a strategy designed to guard the nation and its infrastructure from dangerous people and material, while mitigating the consequences of disasters by strengthening the nation's emergency preparedness and response systems.[2]

But the National Security Agency (NSA), Central Intelligence Agency (CIA), and the FBI all remained outside of the domain of the DHS. Instead, the Coast Guard, the Secret Service, the Border Patrol, and the Transportation Security Administration (TSA) served to provide manpower and military assets for DHS. This dilemma—that DHS was charged with a mission requiring military force, but equipped with no military assets—left the issue of orientation unresolved: was the DHS more a military instrument or a civilian one?

Of course, there were starts and stops to what was the largest restructuring of the federal government since the National Security Act of 1947 and the creation, among other things, of the National Security Council (NSC), the President's advisory body on foreign affairs, and the CIA. Chertoff, who took over from Ridge on February 15, 2005, outlined a six-point review looking to "transform" the department.[3] Chertoff's review was, essentially, an admission of initial failure, which necessitated transformation of the DHS. But what came next was more circumspect in concept and implementation, and perhaps did not represent a transformation so much as a restructuring that ultimately changed little and perpetuated existing problems and concerns.

In Chertoff's view, the biggest change was his insistence on a more powerful administrative arm answering to the director. One part, the Homeland Security Advisory Council, looks to better share information with federal agencies and state and local authorities. The other, the National Infrastructure Advisory Council, provides guidance on securing information systems of public and private institutions critical to the nation's economy. The need to pull the threads of domestic security together, and do so to gain actionable intelligence but not overstep mandates, fell to one individual. Since the management office that had done much of this in the previous organizational scheme had been abolished, the burden was all the greater and raised questions of feasibility and workability of the department at the top. In short, were two advisory bodies really better than one?

Additionally, instead of five directorates, a number of "primary components" (and at times labeled directorates) now constituted the department: Directorate for Preparedness; Directorate for Science and Technology; Office of Intelligence and Analysis; Office of Operations Coordination; Policy Directorate; Domestic Nuclear Detection Office; Federal Emergency Management Agency (FEMA); and Border Security. To some degree, these changes retained the narrow standing (i.e., to fight terrorism) of the DHS when first stood up. But to mitigate the effects of natural disasters, the structure needed to reflect this mission as well. Better policy coordination, better intelligence and information sharing, efficiency of operations, consolidating preparedness assets—all these aims were to be better achieved via reorganization. Of course, it is arguable that these measures could have occurred within the existing structure and had, in fact, been goals of the first organization.

In admitting, however tacitly, this structural flaw, the subsequent reorganization of DHS emphasized investment in the training of personnel so that they would perform better on the job. Linking this step to successful reorganization was specious, however. More concrete, and perhaps more important, were two changes: moving FEMA into a line that reported directly to the Secretary of Homeland Security and creating a new assistant secretary of cyber security to identify and assess cyber threats. These were changes for the better. But these moves, when taken together, highlighted a department establishing its foundation along two different fault lines. One was a move to advance DHS beyond its initial concern of defeating terrorism at home in order to include mitigating the impact of natural disasters, and the second was a recognition of the prominent role played by technology in stopping terror threats to America.

Did these ends require a complete restructuring of the DHS to succeed and did this occur? On paper, a new series of directorates were enacted constituting a transformation. But a paper model only underscored the need for outstanding personnel, and this was, at times, overtly acknowledged. In sum, what would make DHS function best was a streamlined organization staffed with talented and dedicated personnel. This remained the mission under Chertoff as much as it had been under Ridge, so exactly what was transformed remained unclear.

A focus on achieving better security again raised the fault-line issue, however. Renewed emphasis on strengthening the Office of Intelligence and Analysis provided DHS with its own intelligence arm, a near military function. A Directorate for Preparedness, however, brought to the foreground the civilian side of DHS as first responder training, citizen awareness, public health, and infrastructure security were all established as priorities. These steps did not alleviate the tension of defining homeland security as confronting the new threat of terrorism at home and simultaneously addressing the long-standing need to better cope with natural disasters. Furthermore, many experts considered both of these essentially military problems, but the existing DHS structure was disallowed a military component to meet these challenges. This contradiction remains unresolved.

There are additional aspects of the undertaking. DHS represented a consolidation of bureaucracy as a number of existing agencies were simply moved under its umbrella (e.g., transportation, treasury, and agriculture). Others were eliminated and valued features subsumed into other parts of the DHS apparatus (e.g., the Office for Domestic Preparedness and the Energy Security and Assurance Program).[4] For those decrying its size, some 187,000 employees, this centralization was a mitigating factor.

The organizational structures of DHS as described in both incarnations, are, in reality, but a thin layer of bureaucracy trying to pull together existing government agencies that failed to coordinate with one another in the past. With the establishment of DHS, the government got marginally bigger, but arguably more coordinated, at least on paper. The Directorate of Policy, for example, targeted this end directly by ensuring implementation of policies, regulations, and initiatives across DHS. Reality, though, fell to the human element. Could personnel satisfactorily staff the DHS and make existing parts work as a whole? Janet Napolitano, the third head of the department, tacitly endorsed this assumption once again when she did little to change its structure after taking over on January 20, 2009.[5]

Another tension in DHS structure, one which was not resolved but managed, was that of lateral enmeshment in the overall machinery of the U.S. Government. Under Chertoff, the Office of Legislative and Intergovernmental Affairs was charged with the task of ensuring that DHS coordinated its actions with the office of the President and with state and city governments. But the immense need for coordination in this respect made it clear that "homeland security" was a task larger than the already huge department. The reality was that as the larger mandate necessitated ever deeper ties between DHS and the outside world, this fact could be managed but not resolved or eliminated. Access to the President had been accomplished early on with a cabinet seat for the department head. But to subsume state and local authorities into the department was absurd; to name additional figures to oversee such interaction was equally undesirable—adding personnel without any appreciable benefit. The expectation was that dynamic leaders in already existing key positions within DHS could overcome, by successful management, any possible problems of coordination or communication.

GOING FORWARD

Following the structural evolution of DHS, however, does not necessarily speak to its effectiveness. It probably has better coordinated existing governmental agencies. But the more concrete measure of results lies in the answer to more obvious questions: has it helped prevent a terrorist attack, and did it speed a response to a domestic crisis? When asking these questions, the proverbial dilemma becomes how to prove a negative. There has not been a successful terrorist attack on U.S. soil since 9/11, but many experts are not willing to allow DHS to take the credit. The color-coded threat level advisory system (now three colors, not five) came under criticism for heightening concern, but offering no coping methods should the concerns prove validated. Estimates indicate that far too much of U.S. freight, shipping, and other ports of entry are not sufficiently checked. Airport security has called itself into question with search criteria of passengers. In terms of disaster relief, there has not been a crisis on the scale of Hurricane Katrina, and hopefully there will not be one. The British Petroleum (BP) oil spill in the Gulf of Mexico in 2010 exemplified the difficulty of classifying "disasters." Whatever it was, many critiqued the Barack Obama administration for failing to respond quickly and decisively to the problem. Was this the fault of the DHS as well?

What this partial list points to is a department still in need of proving itself. But how can it? Homeland security in the United States is a zero-sum game, and this expectation is given life by domestic politics. A terrorist attack would adversely influence the voters' view of the sitting political party. The BP oil spill probably did have a negative impact on the polls for Democrats in that year, although it would have been but one factor of defeat.

Regrettably, this dynamic brings the conversation full circle—is the fundamental challenge for DHS one of structure or mandate? Can DHS keep the country safe whatever its structure? This chapter has suggested that it cannot. Yet this does not mean the department is dysfunctional, that finding the right organization would solve the problem, or that DHS represents a vain effort at achieving better homeland security. Rather, the preceding analysis suggests that the mandate is simply larger than DHS's capabilities. The question that remains is whether DHS could stop a terrorist attack or respond to a natural disaster such as Hurricane Katrina without a kinetic military component. This question speaks to a larger concern: that the American government still lacks coordination when looking to respond to the challenge of homeland security. The DHS has attempted to marshal the civilian sector. Is this enough?

A NOTE ON SOURCES

There are surprisingly few publications covering DHS. A larger vein of literature addresses the general concept of homeland security. See, for example, *Understanding Homeland Security* by John B. Noftsinger,[6] and Bruce Maxwell, editor of *Homeland Security: A Documentary History.*[7] Often, these studies, like the Maxwell book, present a historical overview and then bring things current. Only a few of these studies comment on the structure of the DHS at some length. Tom Lansford, in *To Protect and Defend: US Homeland Security Policy*, offers a chapter on the department's organization.[8] Jane A. Bullock, in *Introduction to Homeland Security*, provides more comprehensive coverage than does Lansford's book, but again, like Lansford, offers only one chapter, "Organizational Actions," on structure and formation of the department.[9] The overall focus of these two books, like all of this literature, is the general concept of homeland security. Additionally, both books are stilted toward being noninterpretative, the Bullock book in particular, given that it is basically a textbook. So structure and organization of the DHS is present in the most generic and base way imaginable when it is present.

Ridge and Chertoff wrote books addressing their roles in homeland security. Interestingly, neither comments directly on the DHS in any way. Rather, they offer their views on achieving better homeland security, assuming that this is something they imprinted on the organization they led. Both offer shallow analysis, conjecture, and anecdotal evidence.[10]

In sum, there is no good history of the DHS, let alone a good history of its organization and its "stand up" process. Nor is there a historian assigned to the DHS. The U.S. Army Historical Directory 2010 lists the position of "Chief Historian" as vacant. Phone calls to DHS Office of Public Affairs, as listed in the Army directory, produced no one answering to the title of historian in any way. The previous historian was Priscilla Jones, and a phone conversation with her confirmed that position is currently vacant. Jones served under Chertoff and he, according to her, did not value a historian as part of DHS.

More promising is primary research that follows the evolution of the DHS from President George Bush's proposal, to the DHS as conceived of in both houses of Congress, to its fate in conference prior to becoming officially operational. Chertoff's "transformation" of the department has created another extensive paper trail. The steps in the process of "standing up" the DHS are starkly presented in this chapter. It would be valuable to determine the nuance to this picture in terms of the DHS stand up based on this evidence. Also, decisionmakers involved in this process can be identified beyond what is done here, and that is following the lead of the secondary sources and referencing the department heads, Ridge, Chertoff, and in passing, Napolitano.

An extensive look at congressional records, coupled with interviews of key personnel, including the retired former heads of the department, can go a long way to augmenting and confirming many of the important issues and concerns raised in this chapter. Such a chapter promises to initiate the first real substantial book-length history of DHS, a worthwhile endeavor in its own right. Cyber Command would be wise to back such a study, given the DHS's own interest in stopping terrorism and doing so in the cyber world. Avoiding mistakes, emulating successes, and, better still, avoiding duplication of effort, these steps forward could be better undertaken with such a study in hand.

ENDNOTES - CHAPTER 14

1. Steve Fainaru and James V. Grimaldi, "FBI Knew Terrorists Were Using Flight Schools," *The Washington Post*, September 23, 2001, p. A24.

2. Michael Chertoff, *Homeland Security: Assessing the First Five Years*, Philadelphia, PA: University of Pennsylvania Press, 2009, p. 1.

3. Tom Lansford *et al.*, *To Protect and Defend: US Homeland Security Policy*, Derbyshire, United Kingdom (UK): Ashgate, 2006, p. 86.

4. See the DHS chart titled, "History: Who Became Part of the Department" available from *www.dhs.gov/xabout/history/editorial_0133.shtm*, accessed March 10, 2011, or more recently, *www.dhs.gov/who-joined-dhs*, accessed April 17, 2015.

5. See the DHS organization chart available from *www.dhs.gov/xlibrary/photos/orgchart-web.png*, accessed 25 February 2011.

6. John B. Noftsinger, *Understanding Homeland Security*, London, UK: Palgrave Macmillian, 2007.

7. Bruce Maxwell, ed., *Homeland Security: A Documentary History*, Washington, DC: CQ Press, 2004.

8. Lansford *et al.*, "Structure of Homeland Security," Chapter 4, offers a chapter on the department's organization.

9. Jane A. Bullock *et al.*, "Organizational Actions," *Introduction to Homeland Security*, 2nd Ed., Amsterdam, The Netherlands: Elsevier, 2006, provides more comprehensive coverage than does Lansford's book, but again, like Lansford, offers only one chapter on structure and formation of DHS.

10. See Tom Ridge, *The Test of Our Times: America Under Siege – and How We Can be Safe Again*, New York: Thomas Dunne Books, 2009; and Chertoff, *Homeland Security*.

CONCLUSION

CHAPTER 15

CONCLUSION

Ty Seidule

COMMON ISSUES FOR NEW SECURITY ORGANIZATIONS

Since 1947, most newly created security organizations have had at least one of four tacit missions. The first is to create a single, integrated command separate from the parochial services or other federal agencies. The second is to provide more civilian oversight of military and intelligence organizations. The third mission is to integrate other countries into America's security posture. The final mission is to integrate new technology. In many cases, incorporating new technology into security organizations requires overcoming joint, combined, and interagency problems. While technology, then, does require new organizations, an implied purpose is an attempt to overcome service rivalry for roles and missions. Listed in this book are themes that warrant careful thought as new organizations stand up.

Organizational Rivalry.

> The trouble with organizing a thing is that pretty soon folks get to paying more attention to the organization than to what they're organized for.[1]
>
> Laura Ingalls Wilder

We looked at the creation of many different types of security organizations; interservice or interagency competition played a role in almost all of their creations. The fight for general officer billets, resources, and turf is so ubiquitous that it becomes a natural part of the organizational landscape. Major Seanegan Sculley describes in the creation of Joint Forces Command (JFCOM) that a new organization inherently upsets the old order.[2] The Navy never approved of JFCOM despite overseeing it, and throughout the many incarnations of this organization, the Navy worked to undermine it. The reorganization after the September 11, 2001 (9/11) terrorist attacks stripped JFCOM of its operational responsibilities, hastening its death. Faced with reorganization, each service will see opportunity for gains or losses and fight ruthlessly (although bureaucratically, thankfully) for its best interest.

If interservice fighting is inherent, are there remedies? Each reform was to some degree an attempt to quell service rivalry, thereby increasing efficiency and efficacy. Yet, perhaps ironically, the process for creating new organizations has often invited or initiated rivalry, even as the ultimate goal was to minimize it. Simply understanding that service rivalry is a natural byproduct of forming a new organization can lower tension. Most new security organizations have one or two branches championing reform, while at least one sees a threat. As Major Samuel Cook wrote, when U.S. Space Command (SPACECOM) formed, the Army enthusiastically participated because it saw an opportunity to exploit a new medium of warfare—and receive additional funding. On the other hand, the Navy resisted because it worried about losing autonomy and funding.[3] Managing expectations can help. By acknowledging service concerns, leaders can try to lower parochialism through extensive communications with the services. Leadership is the key. However, when resources, prestige, and commands are in the balance, interservice rivalry is devilishly hard to avoid.

Leaders set the tone for interservice interactions, and can just as easily exacerbate service rivalries. When U.S. Central Command (CENTCOM) was formed, the senior military officer for the Army, General Volney Warner, and the Marines, Lieutenant General P. X. Kelly, disagreed over the vision of the new organization. Warner later recognized the extreme tension between the two, as he wrote, "Unfortunately, we were both caught up in the service argument as to whether it should become a premier Army or Marine force."[4] As Major James Harbridge argued in his study of CENTCOM, the atmosphere became so toxic that it affected the staffs and contributed to an unhealthy command climate.[5] Senior leaders in a new organization must work tirelessly to promote a common vision. Subordinates can sense acrimony among their leaders.

Looking to the future, rivalry will continue, but service jealousies may pale beside interagency ones. The Departments of State (State) and Defense (DoD) have always engaged in turf wars, but State has historically (and relatively) had so few resources that the fight was one of symbolism and prestige rather than a fight over power. However, the intelligence fight looks to be much stiffer because the resources and mission of the Director of National Intelligence, the Central Intelligence Agency (CIA), and DoD are overlapping and unclear. The services' fight over roles and missions, which resulted in the "Revolt of the Admirals" in 1949 followed by the compromise "Key West Agreement," was organizationally bloody. It took several laws to end the turf fight. The creation of the Air Force and a disruptive new technology—nuclear weapons—required years to sort out. One could argue that the combination of the 9/11 attack, the creation of the Department of Homeland Security (DHS) and a new disruptive technology (cyber warfare) could lead to another turf war.

Working with Allies.

One element that tends to dampen service rivalry is multinational participation. Since World War II, the United States has participated in the creation of many organizations that combine different nationalities to promote collective security. Major Joseph Scott examines the creation of the North American Aerospace Defense Command (NORAD), in which American and Canadian air force officers worked together to counter the Soviet air threat.[6] The initial contacts were informal and pragmatic. Military officers led the integration process. What began as informal working groups evolved into a bilateral agreement by both legislatures, despite fears on both sides of the border. It remains one of the great multinational organizational success stories in military history.

NORAD's sense of self is apparent in their wildly successful "Santa Claus Tracker," which started in 1955. The mission to track Santa's journey lifted the children's story into the age of technology. Today, NORAD's officers explain that they use "ultra-cool, high-tech, high-speed digital cameras," as well as fighter jets and satellites to track the jolly old elf. Tracking Santa started by mistake. A Colorado Springs, CO, newspaper ad asked local children to call Santa at a department store. The number was typed incorrectly and the calls went to NORAD. The duty officers and noncommissioned officers answered every call, and a tradition was born.

Almost 60 years later, the program takes almost 100,000 calls and e-mails answered by citizens and luminaries such as First Lady Michelle Obama. Corporate sponsors such as Google, Verizon, and Booz Allen Hamilton gladly underwrite financial costs. The Santa Tracker is more than a public relations ploy. Canadian Forces Lieutenant General Marcel Duval, the deputy commander of NORAD, explained that the Santa Tracker helps define the organization. "It's really ingrained in the NORAD psyche and culture. It's a goodwill gesture from all of us, on our time off, to all the kids on the planet."[7] NORAD stumbled into the Santa Tracker, but their organizational culture was open to Santa, in part, because of their multinational mission.

While NORAD's charter is explicitly combined, most American security organizations have an implicit multinational role. Some, like Special Operations Command (SOCOM), with its Foreign Internal Defense mission to organize, train, advise and assist host nation military forces, have a clear assignment to work with other states.[8] Other commands use military-to-military contacts to promote security. Every organization must promote security with other nations, whether they are like-minded democracies or strategic allies. The implied mission of each new security organization is to forge partnerships abroad. An added benefit comes from working with allies—reduced American service rivalry.

American military leaders play a crucial role in selling multinational organizations. The best leadership example among the chapters in this book comes from Dwight Eisenhower. When he became the first Supreme Headquarters Allied Powers Europe (SHAPE) commander, he spent a year on the road travelling to every nation, meeting with politicians, the press, and military leaders to sell the importance of the North Atlantic Treaty Organization (NATO) to a post-war Europe. Leadership engaged outside the organization paved the way for trust and cooperation, both within the organization and external to it.

The Analogy and Metaphor Problem.

> Though analogy is often misleading, it is the least misleading thing we have.[9]
>
> Samuel Butler

As organizations stand up, they look to the past for successful examples. Humans want analogies or models because they help us organize the overwhelming complexity of the present and future in a coherent way. Organizations—made up, of course, of human actors—also look to the past for models. Yet, more often than not, analogies provide a siren song that lead new organizations into trouble. Too often, leaders look only to successful examples and forego the hard analysis necessary to see the differences between the past and current circumstances. Conversely, they ignore examples of past failures and therefore fail to see congruencies in those circumstances. Our natural desire for analogies clouds our judgment and reduces the complexity of current problems to the haze of the past. Even so, the lure of historical analogies is powerful, and ignoring past experience entirely seems also unwise; rather, leaders would do well to use historical examples critically and carefully instead of cherry-picking positive examples to support an already determined conclusion.

Colonel Gregory Daddis provides one example in the Vietnam War in the history of the Office of Civil Operations and Revolutionary Development Support (CORDS), an interagency organization created to coordinate all military and civilian pacification efforts in Vietnam.[10] The American leaders of CORDS found the analogy of the British effort in Malaya an all-too-tempting example of a successful counterinsurgency campaign against a communist enemy. Yet the differences between the two countries and wars were staggering. The British problem occurred on a small peninsula under very different circumstances. For one, the British had already agreed to leave, and unlike in Vietnam, where the Americans also faced a conventional foe in the North Vietnamese Army, the British military did not face a conventional enemy in addition to the insurgents. The leaders of CORDS believed, however, that the Malayan counterinsurgency demonstrated that an organization dedicated to the pacification effort could win the war. The analogy helped to shape the view that better organization and better pacification was the road to victory, ignoring a fundamental difference between the two wars: the South Vietnamese government suffered from terminal internal rot. CORDS was helpless to change an entire foreign government.

Today, the model of choice for new security organizations is SOCOM. SOCOM has had very public successes over the last decade. One indelible image stands out. A bearded Special Forces Soldier wearing Afghan garb directs bombers and drones to defeat the Taliban. He is adaptive, well-trained, and technologically competent—a mule-driving soldier with a satellite phone who captured the imagination of the nation. Inheritors of a maverick or cowboy American tradition, in the public imagination (and, to some extent, in reality), Special Forces Soldiers work in small teams with little guidance, far from higher headquarters. A handful of Green Berets can turn a motley bunch of gun-toting thugs into an army, and its SEALS can take down America's enemies in stealth raids.

Military and civilian leaders see other, more important, reasons to emulate SOCOM. It has a substantial, independent funding stream directly from Congress and a separate personnel system. Every new organization wants to replicate its success—and its money. Moreover, each soldier is better trained, better equipped, and better paid (with large retention bonuses) than their counterparts. Regular military forces look to SOCOM jealously. Special Operations Forces have no support function that takes them away from training. No mundane garrison tasks for them. Deployments are short compared to those of a line infantry unit, which allows them to train more often. Finally, the training for each Special Forces Soldier takes years and includes dedicated language education. No wonder they are the envy of anyone who wears a uniform.

Yet, as Major Brian Dunn writes, Special Operations Command's creation was most unusual and unlikely to be replicated.[11] All four services aggressively tried to stop SOCOM's formation at every turn over the course of 30 years (see service rivalries and turf wars already mentioned). That very intransigence led powerful members of Congress to create an organization despite, and even to spite, the services. Legislators hoped to headquarter the new command near Washington, DC, where key senators and congressmen could oversee their brainchild. Instead, the Pentagon spitefully relegated the headquarters to the

Florida boondocks. After SOCOM's formation, the Army, Navy and Air Force tried to undercut it for years. The Marine Corps refused to participate at all. SOCOM's success came because Congress distrusted the Pentagon and, therefore, lavished SOCOM with designated resources over the long term. No other organization began with so little military input, and no new organization is likely to receive the special treatment SOCOM received and continues to receive from Congress.

While explicitly using analogies is generally problematic, each new organization does have a history. As Major Josiah Grover writes, the multinational security organization SHAPE was able to see the World War II-era SHAEF (Supreme Headquarters Allied Expeditionary Forces) as a starting point but quickly realized the difference between a wartime coalition and a peacetime alliance, and then worked to account for the differences.[12] The leaders of SHAPE understood the past, but saw present issues clearly enough to remain unshackled from analogies.

Seeing a new organization as standing outside history is as problematic as using unsuitable analogies. As Colonel Gian Gentile describes, the U.S. Air Force argued during its formation in 1947 that its technological novelty rendered all previous war irrelevant.[13] Airpower, its adherents claimed, would win war by itself without the need for a punishing ground campaign. Denying the relevance of history, air leaders instead invented a metaphor for the modern nation—the industrial web. Like a spider's web, if a nation's industrial web was damaged at crucial points, the entire web (and therefore the nation) would collapse. The metaphor, like the analogy, took the complex and unpredictable activity of war and simplified it into a manageable concept. Nothing about war or security is simple. While airpower was, and is, a crucial element to warfare, it is not the only one. To date, no technology has either eliminated warfare or caused a nation-state to surrender on its own. New organizations would do well to place themselves within history, not outside it.

If analogies and metaphors are so powerful, what should be done? Curiously, the remedy for over-reliance on historical analogies may, in fact, be to use more of them. If leaders are thinking about multiple analogies, a single model will less likely stick. More examples require people to think critically about what aspects of an analogy are similar and which are different. In *Thinking in Time: The Uses of History for Decision Makers,* historians Richard Neustadt and Ernest May recommend that every time staff officers use an analogy, they take a piece of paper and create two columns, "likenesses" and "differences." When an analogy comes up, using this simple exercise will at least help find the negative aspects of the comparison as well as the positive. Neustadt and May also have a recommendation for putting the problem in historical context to avoid the error made by those who created the Air Force in 1947. Create a timeline for the organization that acknowledges all the previous and current organizations tasked with a similar mission. By putting a new organization on a timeline, staff officers can see their organization on a continuum. This is important because new organizations, by design, will overlap with old organizations.[14]

Simulations.

When a headquarters stands up, the leadership naturally wants to test the unit. How will the new staff react to a simulated wartime situation? What problems can be determined in peacetime? Far better to learn about internal problems in training than during the stark reality of war, the thinking goes. One of the themes that emerges from this book is the crucial role simulations and command post exercises play in the early stages of a new security organization. Leaders craft these exercises carefully for a variety of reasons. As we looked at the various organizations, these early exercises helped create organizations and, in one case, set the stage for a command's death.

Prior to the creation of U.S. Transportation Command (USTRANSCOM), leading logisticians doubted its ability to project and sustain forces should the President call on the Armed Forces to fight a large war. To highlight the problem, general officers created a simulation exercise called NIFTY NUGGET in 1978. As Colonel Gail Yoshitani describes in her chapter, the simulation showed that in a war between the United States and the Union of Soviet Socialist Republics (USSR) in Europe, 400,000 soldiers in theater would die because they failed to receive basic supplies. The post-exercise report argued strenuously for a single command to supervise mobilization and deployment. A well-designed simulation, run honestly and publicly, showed the potential for catastrophic failure without the real casualties of war. Politicians and senior military leaders saw the results and acted, creating what would become USTRANSCOM.

At the other end of the spectrum, Joint Forces Command (JFCOM) executed the largest simulation and field exercise in American history. Dubbed MILLENNIUM CHALLENGE 2002, it tested joint concepts created by JFCOM. As Major Sculley writes, it involved over 10,000 people on 26 different installations and cost in excess of $250 million.[15] Several vocal critics thought the experiment was rigged to validate assumptions based on a doctrine known as Effects Based Operations that is now widely discredited.[16] When the critics found no outlet within JFCOM, they went to the press and Congress. The exercise became a validation of the organization itself, not its doctrine or ability. The perception of both the exercise and the command became one of intellectual dishonesty. When Secretary of Defense Robert Gates was looking for a four-star command to eliminate, he found only a few proponents wishing to keep JFCOM. Unsurprisingly, they were congressional leaders from the command's home state of Virginia worried about jobs. A poorly conceived, expensive, and rigged simulation helped kill its parent organization.

Simulations can be crucial. Properly designed exercises that are well-publicized and seen as fair and accurate can help a fledgling organization. SOCOM, TRANSCOM, and JFCOM wrote about their simulations in journals and other professional outlets. Congressional appropriators and defense officials take simulations seriously as long as they are not seen to be gaming the system in a search of more resources. Leadership is again crucial to ensure staffs understand that the simulation's purpose is an honest assessment of capabilities and doctrine.

Reacting to Failure.

War remains the decisive human failure.[17]

John Kenneth Galbraith

Speaking to West Point's Corps of Cadets in February 2011, Secretary of Defense Robert Gates remarked "that when it comes to predicting the nature and location of our next military engagements, since Vietnam, our record has been perfect. We have never once gotten it right."[18] As a long-time CIA analyst and a historian, Gates understands that the future will always remain unpredictable and dangerous. Despite the billions of dollars America spends annually on intelligence in dozens of agencies, world events have a near-universal ability to surprise even the brightest military and civilian minds. The unknowable questions include where, when, and against whom we will next fight. With hindsight, the attack may seem obvious, but only after the fact. Security organizations must understand that world events will radically change their mission.

Several organizations in this book owe their existence to failures of intelligence collection or processing. As Major Kevin Scott writes, the National Security Agency was one such outfit.[19] In June, 1950, North Korea attacked South Korea, leading to American and United Nations (UN) military involvement. Six months later, Chinese troops crossed the Yalu River. America's intelligence collection agencies missed both attacks. To remedy the problem, the existing defense organizations created in the 1947 National Defense Act were strengthened, forming the National Security Agency and adding thousands of billets to its end strength.

Then, as Professor Matthew Flynn writes, DHS was also a byproduct of an attack. To atone for the intelligence lapses that led to the 9/11 attack, the American people and their government did not demand bureaucratic seppuku from any single agency, politician, analyst, or military leader. Instead, the demand was for "reform." Harking back to the Progressive Era again, reform, many thought, would solve the problem of bureaucratic selfishness in intelligence gathering. Many government leaders argued that, if government agencies had communicated effectively, they could have thwarted the al-Qaeda attack. A better chain of command and more effective interagency processes were the remedies. In this case, civilian agencies would receive the same legislative medicine imposed upon the defense establishment in 1947 and 1986.

The solution was more centralization and unity of command, yet new organizational charts and the creation of the largest federal agency since the creation of the Department of Defense (DoD) in 1947 still failed to force an alphabet soup of military and civilian intelligence agencies to work together. Congress and the President again tried to reform humans through management and bureaucratic efficiency. As Professor Flynn argues, DHS may have crossed the threshold of organizational utility.[20] Could something so big and with so many different missions work effectively? Initially, the answer was no. Two years after its creation, the second director admitted failure and issued a plan to "transform" DHS.

No organization can prepare for mission failure. However, through simulations, exercises and "red cells," a new organization can think through the effect of an attack. The likelihood of an assault using cyber technology seems increasingly likely. Even if a cyber attack is less catastrophic than 9/11 or Pearl Harbor, HI, any disruption will have consequences. The 1983 invasion of Grenada accomplished its mission of saving American medical students and ousting a communist leader linked to Cuba. Nonetheless, the military problems garnered much more interest than the fleeting success. Should failure occur, the government will likely not seek individual accountability. Instead, America's elected representatives will look for organizational problems followed by reform legislation. New organizations, especially those focused on new capabilities, should have suggestions ready for changes in legislation, personnel, and oversight should an attack occur.

Culture.

Each multinational or joint command takes on a personality or culture created by early leaders and decisions. While the culture might be hard to describe by someone inside the organization, visitors, or new arrivals can sense it quickly. Once set, organizational cultures become entrenched and difficult to change. As the cultural anthropologist Mary Douglas has written, 'If you want to change the culture, you will have to start by changing the organization."[21]

Harbridge writes that CENTCOM developed a culture of inferiority because it was seen as a less important regional command.[22] European Command and Pacific Command were older, more prestigious, and had permanently-assigned forces. Moreover, CENTCOM's first commander was only a lieutenant general. These factors made CENTCOM the least among equals, creating a culture that affected its war plans and eventually the way it fought. As a strategic headquarters, it should have designated a subordinate command to fight the operational level of war. Yet, even as the region gained significance within the arena of American security, the persistence of this culture of inferiority led successive CENTCOM commanders to fight at the operational level, including in two invasions of Iraq. CENTCOM needed to focus on the regional implications of war, but that was not the organizational culture. By focusing on the operational level, CENTCOM commanders failed to think about the policy goals when war ended.

"The Best Defense Is a Good Offense."

Starting in World War II, security organizations played a far greater role in American life than they had before. The Pearl Harbor attack created fear and fear caused panic. Panic led to overreaction and a search for fifth column agents in America. In 1942, President Franklin Roosevelt ordered the military to intern hundreds of thousands of Japanese-American citizens in concentration camps. Democracies should worry about the excesses born out of fear.

The threat of carrier-based planes attacking American cities was frightening, but the Cold War era was even worse. Starting in 1949, the threat posed by planes laden with nuclear bombs followed by the even greater existential threat of nuclear tipped intercontinental ballistic missiles induced panic and required further governmental action into American life. (Think, for example, of the "Duck and Cover" cartoons played at schools across the country to teach children how to survive a nuclear blast.) As in World War II, fear led to such pernicious policies as the witch hunt for Communists in the State Department, Hollywood, and even the U.S. Army.

The Armed Forces reacted by creating tens of thousands of nuclear weapons from Atomic Demolition Munitions (tiny backpack nukes) to 15 megaton bombs that could destroy any city. Nuclear weapons on constant airborne patrol were dangerous for Americans as well as for Soviets. Collisions between Strategic Air Command bombers caused several bombs to drop to the ground in the United States and allied countries. Luckily, those weapons never exploded, but today there are two missing nuclear weapons—one off the coast of Georgia and one off the coast of Washington state. The cure for the Soviet threat was nearly worse than the threat itself. Today, only a few rusting yellow signs designating fallout shelters remain from those scary times.

The 9/11 terrorist attacks restored fear about their basic security to Americans. Fear revived elements of the national security state that had lessened as the USSR became sclerotic and then died. In some ways, fear has induced panic similar to the early years of the nuclear era. Are there sleeper cells of Islamic terrorists in America? Questions like that hark back to the worst days of the Red Scare.

The threats facing America require organized governmental response. The 9/11 attack was perpetrated against civilians by an ideologically driven enemy. The reaction from the President was to send military force to deny the enemy sanctuary in Afghanistan. While this mission was difficult and dangerous, it followed established lines of command. The President called it war and placed DoD in charge. Military forces led the effort in Afghanistan and guarded the nation's transportation hubs while the Federal Bureau of Investigation (FBI), CIA, Department of Treasury, the Coast Guard, and dozens of other agencies supported DoD's effort. While the Afghan mission remains difficult and problematic, the issue of cyber attacks presents problems just as acute.

While identifying the enemy in a cyber attack is difficult, that is a technical problem of "cat and mouse" that America's brilliant computer scientists will work to solve. The real problem from an organizational perspective is the intended target. If the problem hits DoD computers, the issue is easy to resolve. Cyber Command takes the lead. When Chinese hackers attacked Google, the lead agency was Google.[23] Who else could claim jurisdiction? Was it DoD? DHS? CIA? FBI? Is this a domestic issue or a foreign one?

The Google attack was against a single business and involved a fairly simple response, organizationally. The next, perhaps more sustained, attack could hit Bank of America, Yahoo!, and Nestle as well as Boise, ID; Sao Paulo, Brazil; the U.S. Department of Treasury, and the Italian Carabinieri, simultaneously. An enemy bent on chaos could create a situation that would involve dozens of businesses and agencies from all over the world, leaving each group groping individually for an appropriate response. What is the protocol to allow feuding security organizations entrée to business? How does an agency prepare for an attack against millions of potential targets? Who has responsibility? It is a vexing problem.

Often security organizations see a loss of initiative when a potential enemy has too many tempting targets. The enemy can choose the place and time for an attack and mass at that location, overwhelming any defense. The answer to a defensive problem is to find a public deterrent or an offensive option. Security organizations hate relying on defensive remedies. If history is a guide, security organizations that face new threats will develop an offensive capability to deter the enemy from acting. If deterrence fails, then a reaction that takes the fight to the enemy may deter future aggressors.

When a new technology emerges, security organizations often look first to defensive remedies as an expedient first step before gravitating to an offense capability. In the late-19th century, the initial reaction to British armored warships stationed in Canada and the Caribbean was to build strong coast artillery batteries, yet Congress funded the program only after a war with Spain. In the early part of the 20th century, hundreds of batteries were built, but with that build-up came an offensive capability in battleships, cruisers, and destroyers. Then President Theodore Roosevelt sent the "Great White Fleet" on a world tour to announce that any potential attackers of the U.S. coast should beware.

While defensive capabilities provide some deterrence and make the American people feel safer, defense professionals want an offensive capability. In the nuclear era, America tried both defensive and offensive options. On the defensive side, billions of dollars went to solutions such as the Nike/Hercules missiles meant to destroy a Soviet air attack on American soil as well as the anti-ballistic missile defense of the 1960s and 1970s and the Strategic Defense Initiative (Star Wars). Yet most of the defense budget went to offensive capability, specifically the threat of overwhelming retaliation from a 24-hour manned bomber force by the Air Force's Strategic Air Command and then from intercontinental and sea launched ballistic missiles.

The current war on terror provides another example of how a defensive policy changes to an integrated offensive one. In the 1990s, the Bill Clinton administration reacted to violent Islamic radical attacks in New York and Africa with civil prosecutions and cruise missiles. After the 9/11 attacks, the United States used DoD and the Treasury and Justice Departments, as well as the CIA, to attack al-Qaeda overseas simultaneously and in depth. A declared war allows the government to react in an offensive way.

Perhaps the government's successful response to the al-Qaeda threat provides a model of how to provide defense against cyber attacks as well. When an attack comes, neither a military nor cyber reaction will be sufficient to address an anonymous cyber threat. All agencies in the federal government with assistance from business, state, local, and foreign governments will need to work together to identify and attack the enemy simultaneously and in depth, using all of the same tools that have worked so well against al-Qaeda over the last decade. Good intelligence is the key, and good intelligence comes from a variety of sources that only an interagency and international coalition can provide.

While cyber attacks will continue, now is the time for America's cyber warriors to build relationships with local and foreign governments. As Winston Churchill said, "There is at least one thing worse than fighting with allies—and that is to fight without them."[24] Because cyber attacks easily cross national boundaries, partners will be a requirement to counter the threat. Creating partnerships will be difficult for business and government both in the United States and outside it, but partnerships created in peace will pay dividends in war.

What will be the cyber equivalent of a deterrent or an offensive option? Will offensive action make America safer or less safe? When an attack does come, will fear rule? History provides no easy answers, and neither do we. Smart, dedicated leaders at all levels will make the best decisions they can based on incomplete information. When an attack occurs, which it will, the organization will be ready to fight as well as it can. When the attack is over, it should conduct a thorough review and recommend changes and reforms to statutes, regulations, standard operating procedures, and training.

What history does provide is perspective. America has an excellent record of protecting its citizens. Since the ratification of the U.S. Constitution in 1789, there have been four major attacks on American soil—the British invasion in 1814, the Civil War from 1861 to 1865, the Pearl Harbor attack in 1941, and the 9/11 attacks in New York and Washington. In each case, America reacted with policies meant to protect the American people that also took away the civil liberties of significant parts of the population. Even in times of relative peace, Americans have reacted to perceived threats with coercive policies. Recall, the measures taken during the "red" scares in the 1920s and 1950s came from the threat of attack, not actual attack. Leaders of security organizations would do well to remember the Hippocratic Oath, "First do no harm."

Overall, American politicians and security organizations have an admirable record of balancing security concerns. However, as the investment ads say, "Past results are no indication of future performance." We hope the chapters in this book and the lessons at the end of it will help the reader learn from the successes and the failures of American security organizations over the last 70 years. History provides no answers, but thinking critically about the past can bring wisdom to current security problems.

ENDNOTES - CHAPTER 15

1. Laura Ingalls Wilder, *Little Town on the Prairie*, New York: Harper Collins, 2004, p. 214.

2. Seanegan Sculley, "Sailing on Stormy Seas: U.S. Joint Forces Command and Reorganization in the Post-Cold War World," Chap. 3, this volume.

3. Samuel P. N. Cook, "Commanding the Final Frontier: The Establishment of Unified Space Command," Chap. 6, this volume.

4. Interview with General Volney F. Warner, United States Army Retired, by Dean M. Owen, U.S. Army Military History Institute Senior Officer Oral History Program, U.S. Army War College, Carlisle, PA, 1983.

5. James C. Harbridge, "Organizational Insecurity and the Creation of U.S. Central Command," Chap. 4, this volume.

6. Joseph C. Scott, "Watching the Skies: The Founding of North American Air Defense Command," Chap. 11, this volume.

7. Dan Elliot, MSNBC, December 24, 2010, available from *www.msnbc.msn.com/id/40803123/ns/us_news-life/#.TyG98oF8uuI.*

8. *Special Operations Reference Manual*, MacDill Air Force Base, FL, available from *www.fas.org/irp/agency/dod/socom/sof-ref-2-1/SOFREF_Ch1.htm.*

9. Samuel Butler, Henry Festig Jones, ed., *The Notebooks of Samuel Butler*, New York: Fifield, 1912, p. 94.

10. Gregory A. Daddis, *No Sure Victory: Measuring U.S. Army Effectiveness and Progress in the Vietnam War*, Oxford, MA: Oxford University Press, 2011.

11. Brian P. Dunn, "Legislating Change: The Formation of US Special Operations Command," Chap. 2, this volume.

12. Josiah Grover, "Standing up SHAPE: The Quest for Collective Security in Western Europe," Chap. 10, this volume.

13. Gian P. Gentile, How Effective is Strategic Bombing: Lessons Learned from World War II to Kosovo, New York: New York University Press, 2001.

14. Richard A. Neustadt and Ernest May, *Thinking in Time: The Uses of History for Decision Makers,* New York: Free Press, 1988.

15. Sculley.

16. See H. R. McMaster, "Crack in the Foundation: Defense Transformation and the Underlying Assumption of Dominant Knowledge in Future War," Student Research Paper, Carlisle, PA: U.S. Army War College, 2003, available from *oai.dtic.mil/oai/oai?verb=getRecord&metadataPrefix=html&identifier=ADA416172.*

17. Sculley; John Kenneth Galbraith, *The Economics of Innocent Fraud: Truth for Our Time,* New York: Houghton Mifflin, 2004, p. 62.

18. Robert Gates, Transcript of Speech at West Point, New York, February 25, 2011.

19. Kevin A. Scott, "Trial and Error: The Creation of the National Security Agency," Chap. 13, this volume.

20. Matthew J. Flynn, "The Department of Homeland Security," Chap. 13, this volume.

21. Mary Douglas, quoted in Joseph Demakis, compiled and ed., *The Ultimate Book of Quotations,* Raleigh, NC: Lulu Enterprises, 2012, p. 50.

22. Harbridge.

23. John Markoff, "Cyber Attack on Google Said to Hit Password System," *The New York Times*, April 19, 2010.

24. Warren F. Kimball explains the history of this phrase in "'Fighting with Allies': The Hand-Care and Feeding of the Anglo-American Special Relationship," USAFA Harmon Memorial Lecture #41, U.S. Air Force Academy, CO, 1998, available from *www.usafa.edu/df/dfh/docs/Harmon41.pdf.*

ABOUT THE CONTRIBUTORS

SAMUEL P. N. COOK served as a major in the U.S. Army, finishing as an instructor in the Department of History at the United States Military Academy, West Point, NY, received a B.S. from the United States Military Academy and an M.A. from New York University.

GREGORY A. DADDIS is a colonel in the U.S. Army and serves as an Academy Professor in the Department of History at the United States Military Academy, West Point, NY. He is the Division Chief for American History, and teaches courses in military and American history. He specializes in the history of the Vietnam War, American strategy, and counterinsurgencies. Colonel Daddis's publications include *No Sure Victory: Measuring U.S. Army Effectiveness and Progress in the Vietnam War* (2011) and *Westmoreland's War: Reassessing American Strategy in the Vietnam War* (2014). Colonel Daddis holds a B.S. from the United States Military Academy and an M.A. and Ph.D. from the University of North Carolina at Chapel Hill.

BRIAN P. DUNN is a major in the U.S. Army. He works as a Foreign Area Officer and is currently serving with the Defense Attaché Service in Rwanda. Major Dunn taught in the History Department at the United States Military Academy at West Point, NY, where he taught courses in American history and specialized in contemporary American cultural history. Major Dunn holds a B.A. from Loyola College Baltimore in history and an M.A. in American history from the University of Pennsylvania.

KEVIN W. FARRELL is currently a partner at Cambridge Leadership Development Corp. Dr. Farrell retired as a Colonel in the United States Army, where his last assignment was as an Academy Professor in the Department of History at the United States Military Academy, West Point, NY, where he led both the military and international divisions. Dr. Farrell is a specialist in modern European and German history and is the author of *The Military and the Monarchy: The Case and Career of the Duke of Cambridge in an Age of Reform* (2011). Dr. Farrell holds a B.S. from the United States Military Academy and an M.A., M.Phil., and Ph.D. in history from Columbia University.

MATTHEW J. FLYNN is an Associate Professor of Military History at the Marine Corps Command and Staff College in Quantico, VA. He previously taught at the United States Military Academy, West Point, NY, in the Military and International Divisions of the History Department. He is a specialist in comparative warfare of the United States and the world. Dr. Flynn's book publications include a recent co-authored study titled, *Washington & Napoleon: Leadership in the Age of Revolution* (2012); *First Strike: Preemptive War in Modern History* (2008); and *Contesting History: the Bush Counterinsurgency Legacy in Iraq* (2010). Dr. Flynn received his Ph.D. in history from Ohio University.

GIAN P. GENTILE is a Senior Historian with the RAND Corporation, Santa Monica, CA. He retired as a colonel from the U.S. Army, where his last assignment was as an Academy Professor at the United States Military Academy, West Point, NY. He served as the Division Chief for the Military History and

American History divisions and specialized in military history, American history, and the history of insurgencies and counterinsurgencies. Dr. Gentile is the author of *Wrong Turn: America's Deadly Embrace of Counterinsurgency* (2013) and *How Effective is Strategic Bombing? Lessons Learned From World War II to Kosovo* (2000). Dr. Gentile holds a B.A. from the University of California-Berkeley and an M.A. and Ph.D. in history from Stanford University.

JOSIAH GROVER is a major in the U.S. Army and currently assigned as a Strategist in the G5 shop for the 10th Mountain Division. Major Grover taught in the History Department at the United States Military Academy, West Point, NY, where he taught courses in military history and specialized in American strategy and policy during the Cold War. Major Grover holds a B.A. in history from Providence College and an M.A. in history from the University of North Carolina at Chapel Hill.

JAMES C. HARBRIDGE is a major in the U.S. Army and currently serves as the Executive Officer and Operations Officer (XO/S3) 1st Battalion, 41st Infantry Regiment, 3rd Infantry Brigade Combat Team, 1st Armored Division based at Fort Bliss, TX. He previously served as an Assistant Professor and Deputy Director for the Defense and Strategic Studies program at the United States Military Academy, West Point, NY. Major Harbridge holds a B.A. in history from the Citadel, and a master's degree in international affairs and national security from the Bush School of Government at Texas A&M University.

JOSEPH C. SCOTT is a major in the U.S. Army and currently assigned to the XVIII Airborne Corps. Major Scott previously taught in the Department of History at the United States Military Academy, West Point, NY, where he taught courses in Military History. He specializes in American nuclear policy and strategy in the Cold War and Civil War history. Major Scott holds a B.A. from Dartmouth College and an M.A. in history from the University of Virginia.

KEVIN C. SCOTT is a major in the U.S. Army, where he is currently assigned to the Defense Intelligence Agency, Middle East and Africa Regional Center, where he is the Syria Branch Chief. He previously taught in the Department of History at the United States Military Academy, West Point, NY, where he taught courses in International and Latin American History. Major Scott holds an M.S. in strategic intelligence from the National Intelligence University and an M.A. in European history from the State University of New York-Albany.

SEANEGAN P. SCULLEY is a major in the U.S. Army and is currently serving as the Brigade Operations Officer for the 171st Infantry Brigade at Fort Jackson, SC. Major Sculley taught in the History Department at the United States Military Academy at West Point, NY, where he taught courses in American history and specialized in colonial and early American history. He is the author of *Men of the Meanest Sort: Military Leadership and War in the New England Colonies, 1690-1775* (2008). Major Sculley holds a B.A. in history from Southwest Texas State University and an M.A. in history from the University of Massachusetts-Amherst where he is currently finishing his Ph.D. in early-American history.

TY SEIDULE is Professor and Head of the Department of History at the United States Military Academy at West Point, NY. Commissioned in armor, he has served in the United States, Germany, Italy, the Balkans, and the Middle East. He has also served as the National Security Affairs Visiting Professor at the Naval Postgraduate School in Monterey, California. He is the senior editor of *The West Point History of Warfare*, a 71-chapter military history enhanced e-book featuring animated maps, interactive graphics, 3D images, and primary source accounts. He is the editor of *The West Point History of the Civil War* (2014). Colonel Seidule received a B.A. from Washington and Lee University, and an M.A. and Ph.D. in history from The Ohio State University.

MICHAEL WARNER serves as the Command Historian for U.S. Cyber Command. He has written and lectured widely on intelligence history, theory, and reform, and he teaches as an Adjunct Professor at Johns Hopkins University and American University. His new book, *The Rise and Fall of Intelligence: An International Security History*, was published by Georgetown University Press in 2014. Dr. Warner earned his Ph.D. in history from the University of Chicago.

JACQUELINE E. WHITT is an Associate Professor of Strategy at the Air War College at Maxwell Air Force Base, Montgomery, AL, where she teaches courses in strategy, American Studies, Military History, and Religious Studies. Dr. Whitt previously taught in the History Department at the United States Military Academy at West Point, NY, where she taught courses in Military History and American History. She is the author of *Bringing God to Men: American Military*

Chaplains in the Vietnam War (2014). Dr. Whitt holds a B.A. in history and international studies from Hollins University, and an M.A. and Ph.D. in military history and American history from the University of North Carolina at Chapel Hill.

GAIL E. S. YOSHITANI is a Colonel in the U.S. Army and serves as the Deputy Department Head in the Department of History at the United States Military Academy, West Point, NY, where she has also taught courses in American History and Military History. Her research specialization is in modern American history and U.S. foreign policy. Colonel Yoshitani holds a B.S. from the United States Military Academy, an M.S. from Troy State University, and an M.A. and Ph.D. in history from Duke University.